Corporate Leaders discuss how India can become an Innovation hub at SBAC 2017

National winners of L'oreal's international business competition 'Brandstorm'

SPJIMR wins top honours at Bangkok Business Challenge 2017

Academic Programmes

PGDM

The Post Graduate Diploma in Management (PGDM) is a two year full-time residential programme equivalent to an MBA. The programme is approved by AICTE and accredited by Association of MBAs (AMBA), U.K. PGDM helps build socially sensitive, value-driven leaders with a global mindset.

Placement highlights 2017

Average Package - Rs. 20.9 lpa

Highest Package - Rs. 39.5 lpa

- More than 90% of the batch has offers of Rs. 15 lakhs and more
- The median salary for the batch of 229 students was Rs. 20 lakhs

Prominent recruiters:

TAS, Deloitte, Accenture Strategy, Mahindra GMC, HUL, P&G, Microsoft, PwC-PRTM, Axis Bank, JM Financial, Goldman Sachs, Reckitt Benckiser, General Mills, ITC, Johnson & Johnson, GEP, Miebach Consulting, Samara Capital, Investec, DBS Bank.

PGPM

The Post Graduate Program in Management (PGPM) is a one-year full-time residential programme approved by (AICTE) and is accredited by the 'Association of MBAs (AMBA), U.K., which upholds the highest international standards of management education.

Placement highlights 2017

- Average Package : Rs.17.53 lpa
- Highest Package : Rs.35.42 lpa
- 81% of participants have been placed in Business Consulting& IT&ITES firms.
- 16% of participants have been placed in Manufacturing / Supply Chain Logistics

Prominent recruiters:

Microsoft, Fidelity, Amazon, PWC, SAP, Syntel, SBI Cap, Freshdesk, Mphasis, Shell, KPMG, Accenture, Forrester, EY, PWC, KPMG, Deloitte.

GMP

Choose to study for an international MBA or an MSc. at one of the following four European B-Schools in partnership with SPJIMR under the Global Management Programme (GMP).

- ESB Business School, Reutlingen, Germany
- IESEG School of Management, Paris, France
- Nyenrode Business Universiteit, Breukelen, Netherlands
- EBS Business School, Oestich-Winkel, Germany

Placement highlights 2017

- Average starting salary for an MSc is 38,500 Euros and for an MBA 45,000 Euros.

Prominent recruiters:

World Bank, ING Bank, ABN AMRO Bank, Apple, Airbus Helicopters, Amazon EU, Credit Suisse, E.ON, PWC, E&Y, Deloitte, Crane Co., Capillary Technologies.

www.facebook.com/SPJIMR | https://www.linkedin.com/school/215708 | https://twitter.com/SPJIMR

Address: Munshi Nagar | Dadabhai Road,Andheri West | Mumbai - 400 058, India

Tel: +91 22 26230396 / 022 26232401 |

www.spjimr.org

SPJIMR is a part of the Bharatiya Vidya Bhavan and operates out of its one and only campus at Munshi Nagar, Dadabhai Road, Andheri West, Mumbai. The Institute has currently no branches or campuses in any other city or country.

DATA INTERPRETATION

Practice Book for the CAT and Other MBA Entrance Examinations

Nishit K. Sinha

Pearson

To the one who joined this universe "Riding a Horse"

Editor–Acquisitions: Sharel Simon
Assistant Editor–Production: Karamjeet Singh Khosa

ISBN: 978-93-325-1941-1

First Impression, 2014
Second Impression, 2017
Third Impression

Published by Pearson India Education Services Pvt. Ltd, CIN: U72200TN2005PTC057128.
Formerly known as TutorVista Global Pvt. Ltd, licensee of Pearson Education in South Asia.

Head Office: 15th Floor, Tower-B, World Trade Tower, Plot No. 1, Block-C, Sector-16, Noida 201 301, Uttar Pradesh, India
Registered Office: 4th Floor, Software Block, Elnet Software City, TS-140, Block 2 & 9, Rajiv Gandhi Salai, Taramani, Chennai 600 113, TamilNadu, India.
Fax: 080-30461003, Phone: 080-30461060
www.pearson.co.in, Email id: companysecretary.india@pearson.com

Compositor: Tantla Composition Services, Chandigarh
Printer: Sai Printo Pack Pvt. Ltd.

Contents

Acknowledgements

At the onset, let me say that this book has been prepared with sheer hard work and dedication. I have been working on this book for the past one and a half years – selecting data sets, developing questions and writing explanations. There are a number of people I would like to thank for their help and contribution, without which this book would be incomplete.

First of all, I would like to thank all the students at dueNorth Academics who provided me with their feedback on the questions and solutions. Special thanks to Mr Gaurav Arora, Mr Piyush Gumber, Mr Barjeev Singh and Mr Anurag Chauhan.

I am also thankful to my colleagues – Ms Ritika Sharma, Mr Govind Thapa and Mr Arun Kumar who helped me immensely in preparing the manuscript.

I would like to express my heartfelt gratitude to Mr Rajesh Shetty, Mrs Nitkiran Bedi, Mr Karamjeet Singh Khosa and Ms Sharel Simon who gave the final shape to the manuscript and brought this book in its current format.

Nishit K. Sinha

Preface

This book has been prepared keeping in mind the requirements of the students. The content of the book has been planned following a simple concept of three C's – crisp, concise and comprehensive.

The book solves the following purposes:

(a) It discusses the theory of Data Interpretation (DI), several approximation methods practically through questions.
(b) The book prepares you for the D–Day. Data sets have been put in a random order to test your observation, interpretation and calculation skills.
(c) The content would expose the students to the various types of questions that usually appear in the CAT and other MBA entrance examinations. There are some DI sets, which are based on only one type of data set, and there are some, which are a combination of two or more data sets.
(d) The layout has been planned keeping in mind the need to focus on the reaction time. Since the questions of various levels of difficulty have been grouped together, you may find that the second question in a set is easier than the first one, or may be the last question is the easiest one. So, this book puts you in 'the real testing scenario' and trains your brain to identify the question's difficulty level and then proceed.

I sincerely hope that the users of this book will be benefitted immensely after solving all the DI sets given in the book, and will be able to score well in the DI section.

For any constructive feedback, I can be reached at: nsinha.alexander@gmail.com

Happy Learning!

Nishit K. Sinha

How to use this book

Different steps are involved in solving a question set in Data Interpretation (DI). In my opinion, the following are those steps:

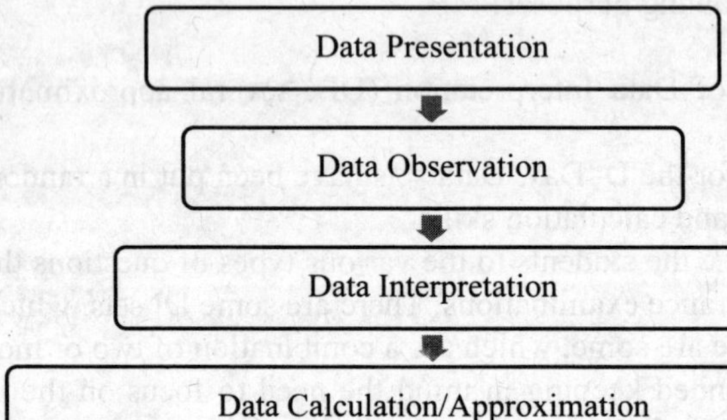

EXPLANATION OF TERMS

Data Presentation

Data collected can be presented in different forms – either pictorial or non-pictorial (meaning through words only). Pictorial dataset can be a table, a bar chart, a line chart, a pie chart, a radar diagram etc. It is quite possible that the same data set can be presented in more than one form. For instance, a line chart can be presented with the help of a bar chart or a table, though it is not necessary that every data set can be presented by each of the other forms. For example, it might not be possible to convert a line chart containing the GDP of two countries over last five years by a single pie chart. Further, there may be a typical way of presenting the data more suitable for a set of data. For example, line chart is mostly used for chronological data. Similarly pie chart is used for presenting the composition of a data set where sum of all the data points is given (either in value terms or percentage terms).

Data Observation

Before we start doing the calculation, it is advisable that you should go through the question to understand what skill is required to solve it? Is it only observation-based, or calculation-based or a combination of both?

There are typical methods of calculation pertaining to different data sets like pie charts or bar charts or line charts etc. In this book we have discussed these methods through questions.

Data Interpretation

Data Interpretation is the act of transforming the given data points into meaningful conclusions. These data points can be words, numbers, pictures, or a combination of more than one.

In CAT examination, you will be asked to solve the questions based upon the given data set.
Before you start solving any set in this book, I would suggest you the following:

(a) Go through the questions first: This will give you a basic idea as to what is to be extracted out of the maze of information provided in words or pictorial format. In fact, you should do this first while solving the questions of DI in any examination.
(b) Once you have attempted the question set, you must go through the solution and see what short cuts can be learnt for the particular data set. In most of the data sets, a storyline has been given, to give you an idea of what was the summary of the set and what questions can be expected in such kind of data sets.
(c) There are questions with four options as well as five options. Solve all the sets.
(d) Ideal number of attempts would be 2 to 3 practice sets a day, investing 10 to 12 minutes per set. Focus more on accuracy than attempting all the questions in a set.
(e) Do not get into the technicalities (the definitions, data and its presentation) of the data set. Your objective should be to solve the questions in those 10 to 12 minutes.

Data Calculation/Approximation

Once you reach the stage of final calculation in any question, I would suggest you to follow the "Thumb Rules of Approximation":

(a) Go through the options: This will give you an idea about the scope of approximation. Ideally, as a thumb rule, if options are at a gap of less than 5 per cent, go through the actual calculation, and DO NOT use approximation.
(b) Your approximation should be in the range of ± 2 per cent. Though I would reiterate that it mostly depends upon point (a) as mentioned above that how much you should approximate the result.
(c) Must know the direction of approximation: You should know if your approximated result is more than the actual result or less than the actual result.

Telecom Operators

Set 1

Direction for questions 1–4

Go through the data set given below and solve the questions based on it.

Following chart gives the revenue of five telecom companies generated in Jan' 2010 and Jan' 2011. All the values are in the multiples of 50.

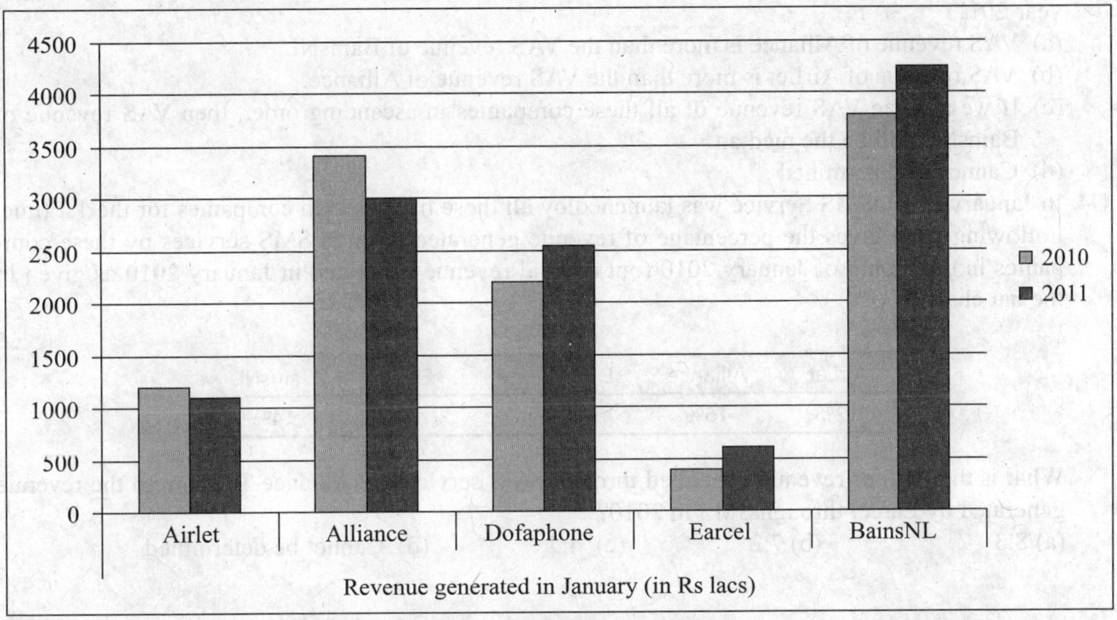

Following table provides the number of users of these telecom companies for the same two years in the month of January (in lacs):

	Number of Users (in lacs)				
	AirLet	Alliance	Dofaphone	Earcel	BainsNL
2010	242	324	286	42	422
2011	259	318	294	46	440

Q1. ARPU is defined as Average Revenue Per User. For which of the following companies ARPU in the month of January maximum for both the years combined?

(a) Alliance (b) Dofaphone (c) Earcel (d) BainsNL

Q2. ARPU is defined as Average Revenue Per User. For how many companies, ARPU has increased in January 2011 over January 2010?

(a) 2 (b) 3 (c) 4 (d) 5

Q3. In January 2011, VAS (Value Added Services) was launched by all these five telecom companies for the first time. Following table gives the percentage of revenue generated through VAS by these companies in the month of January 2011 (out of total revenue generated in January 2011 as given in the bar chart):

AirLet	Alliance	Dofaphone	Earcel	BainsNL
18%	9%	12%	22%	6%

Which of the following is NOT correct for the VAS revenue generated by these companies for the year 2011?

(a) VAS revenue of Alliance is more than the VAS revenue of BainsNL.
(b) VAS revenue of AirLet is more than the VAS revenue of Alliance.
(c) If we arrange VAS revenue of all these companies in ascending order, then VAS revenue of BainsNL will be the median.
(d) Cannot be determined

Q4. In January 2010, SMS Service was launched by all these five telecom companies for the 1st time. Following table gives the percentage of revenue generated through SMS services by these companies in the month of January 2010 (out of total revenue generated in January 2010 as given in the bar chart):

AirLet	Alliance	Dofaphone	Earcel	BainsNL
12%	16%	9%	6%	14%

What is the ratio of revenue generated through SMS services by Alliance Telecom to the revenue generated by Earcel through SMS in 2010?

(a) 8:3 (b) 7:2 (c) 4:7 (d) Cannot be determined

Rough Work

Answer Grid

Q1	Q2	Q3	Q4
c	b	b	d

Solutions to Q1–Q4

A1. Following is the list of ARPU:

	AirLet	Alliance	Dofaphone	Earcel	BainsNL
ARPU Jan 10	4.7	10.49	7.5	10.7	9
ARPU Jan 11	4.2	9.4	8.67	13.04	9.6

It can be seen clearly that Earcel is maximum for both the years. Hence it will be maximum for both years combined also. Hence option (c) is the answer.
Alternatively, it can be observed that all the bars except Earcel gives less than 10 ARPU in the 1st year and in 2nd year, Earcel is maximum.

A2. Following is the list of ARPU:

	AirLet	Alliance	Dofaphone	Earcel	BainsNL
ARPU Jan 10	4.7	10.49	7.5	10.7	9
ARPU Jan 11	4.2	9.4	8.67	13.04	9.6

Hence option (b) is the answer.

A3.

VAS	AirLet	Alliance	Dofaphone	Earcel	BainsNL
2011	18%	9%	12%	22%	6%
Revenue	198	270	306	132	255

Statements (a) and (c) are true and (b) is false. Hence answer is option (b).

A4. Data is given for January 2010, and question is asking for the whole year 2010. Hence cannot be determined. Hence option (d) is the answer.

Average Income and Expenses

Set 2

Direction for questions 1–3

Go through the data set given below and solve the questions based on it.

Following bar charts give the average income and average expense for five families in the five years from 2006 to 2010. Average Savings for the Year = Average Income – Average Expenses.

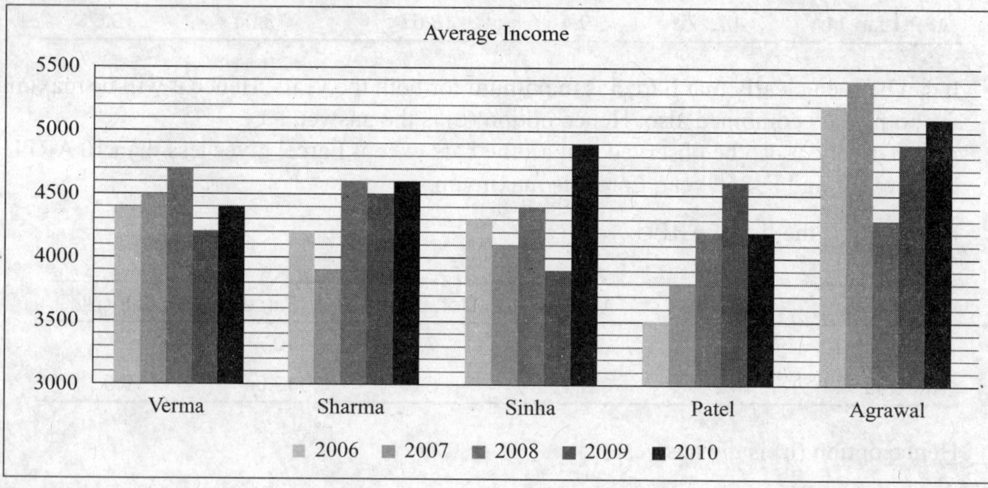

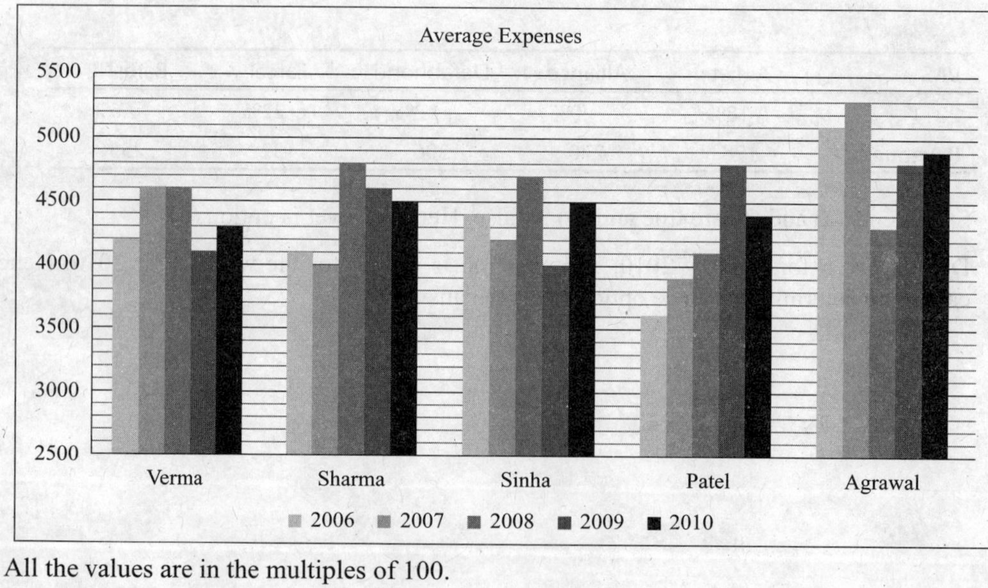

All the values are in the multiples of 100.

Q1. For how many families, savings is non-negative for each of the years?
(a) 0 (b) 1 (c) 2 (d) 3

Q2. Which of the following families' total savings is maximum over the given period?
(a) Agrawal (b) Sharma (c) Patel (d) Verma

Q3. For how many families, total savings is negative over the given period?
(a) 2 (b) 3 (c) 4 (d) 5

Answer Grid

Q1	Q2	Q3
b	a	b

Solutions to Q1–Q3

This question set is purely based upon observation. That also, very careful observation. During online CAT, such kind of questions have been reported by the students where solving a question was more a job of 'eyes' than the job of 'brain'.

Following is the detailed list of savings:

	2006	2007	2008	2009	2010	Summation
Verma	200	–100	100	100	100	400
Sharma	100	–100	–200	–100	100	–200
Sinha	–100	–100	–300	–100	400	–200
Patel	–100	–100	100	–200	–200	–500
Agrawal	100	100	0	100	200	500

A1. For Verma family, savings is negative in 2007, so we do not need to check for other years.
For Sharma Family also, savings is negative in 2007.
For Sinha and Patel family, savings is negative in the first year itself.
For Agrawal family, this is zero for 2008, and positive for all the years. Since question is asking for non negative value, Agrawal family's data satisfy this condition. Hence option (b) is the answer.

A2. Only two families have positive savings—Verma and Agrawal. Agrawal's savings is more than Verma's savings. Hence option (a) is the answer.

A3. Sharma, Sinha and Patel family have negative total savings. Hence option (b) is the answer.

Loan Disbursal

Set 3

Direction for questions 1–3

Go through the data set given below and solve the questions based on it.

Following table gives the loans disbursed (only principal value and NOT the interest) across sectors (in Rs Crore) by a bank in the period 2005–2010. There is no overlapping of loan disbursals across the given sectors. All the Loans disbursed are of only two types – Standard loan or NPA Loan.

Sectors	Standard Loan	NPA Loan
Coal	1481.41	11.27
Mining	4517.06	62.97
Iron and Steel	41972.67	836.9
Metal Products	8196.7	294.08
All Engineering	17903	475.6
Electricity	9052	34.77
Cotton Textiles	16907.7	593.27
Jute Textiles	275.7	17.57
Other Textiles	17264.1	536.9
Sugar	4718.9	26.1
Tea	375.3	102.5
Food Processing	13325.1	590.2

Standard loans are those loans for which repayments are regular and NPA Loan (Non Performing Asset Loan) are those loans for which repayment installments have stopped coming and bank have considered these loa ns as 'bad debt'.

Q1. For how many sectors, NPA loan is more than 2% of the total loan?
(a) 5 (b) 6 (c) 7 (d) None of these

Q2. For which sector, NPA loan as a percentage of total loan the maximum?
(a) Jute Textiles (b) Metal products (c) All Engineering (d) None of these

Q3. Which of the following correctly presents the range of value of standard loan as a percentage of total loan disbursed for all the sectors taken together for the whole period?
(a) Between 96% and 97% (b) Between 97% and 98%
(c) Between 98% and 99% (d) None of these

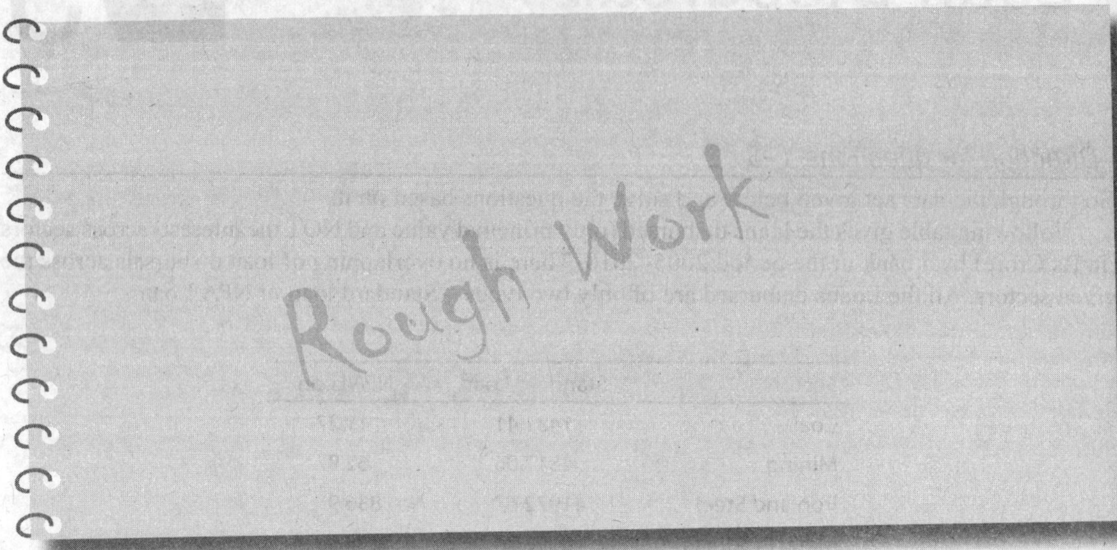

Answer Grid

Q1	Q2	Q3
c	d	b

Solutions to Q1–Q3

A1. Following sectors have more than 2% NPA loan—Metal Products, All Engineering, Cotton Textiles, Jute Textiles, Other Textiles, Tea and Food Processing. Hence option (c) is the answer.

A2. It can be clearly seen that maximum value occurs for Tea (roughly 21%). This question was a test of observation only. Hence option (d) is the answer.

A3. Total standard loan = 135989.6 cr
Total disbursal = 139571.8
Hence percentage = 97.43%
Alternatively, it can be observed that NPA loan as a percentage of total loans comes equal to 2%–3%. Hence standard loan = 100% – (2% to 3%) = 97% – 98%. Hence option (b) is the answer.

McDi Test

Set 4

Direction for questions 1–4

Go through the information given below and solve the questions based on it.

You are the Quality Control Manager at the restaurant McDi and facing a dilemma. Go through the case facts given below and solve the questions based on it:

McDi produces chicken nugget box (1,000 boxes at a time in a batch). McDi has to ensure that it does not supply undercooked nuggets (known as defective) to the customers. If a customer receives a defective nugget box, company has to pay a penalty of Rs 150 per nugget box. To check the level of preparedness of nuggets, any of the two tests can be used:

Test 1		Test 2	
Cost	Rs 6 per nugget box	Cost	Rs 9 per nugget box
Detection Rate	80%	Detection Rate	100%

McDi can re-cook undercooked nuggets at a cost of Rs 75 per nugget box. This box can be now sold to customer as it is now certainly properly cooked. All the boxes which are detected as defective must be re-cooked and after that, these boxes are properly cooked now.

Let 'x' be the number of defective nugget boxes per batch.

Q1. Manager should not use any quality control test, if it is expected that

(a) $x < 100$ (b) $100 < x < 200$ (c) $x > 200$ (d) None of these

Q2. If $200 < x < 400$, then

(a) You should use Test 1 (b) You should use Test 2

(c) You can use Test 1 or Test 2 (d) None of these

Q3. If $x = 1000$, you should use

(a) Test 2 (b) Test 1 and Test 2

(c) Test 1 or No test (d) None of these

Q4. Company should be indifferent between Test 1 and No Test if it finds that the number of defective boxes is equal to

(a) 50 (b) 100 (c) 200 (d) Not possible

Rough Work

Answer Grid

Q1	Q2	Q3	Q4
a	b	a	b

STORYLINE

Total Cost = Cost of Testing + Re-cook Charge + Penalty

It can be seen that if the number of defective nugget boxes per batch is very low, for example, 10, then it is better to pay the penalty on this, than to use any test.

Solutions to Q1–Q4

A1. (a)

Total Cost = Cost of Testing + Re-cook Charge + Penalty
Using Test 1, total cost of testing = Rs 6 × 1000 = Rs 6000
Using options, at x = 100, Cost of Test 1 = Penalty paid for x = 100.
Using Test 1, Detection = 80 % of 100 = 80. These 80 will be reworked upon and for remaining 20, company will pay penalty.
Total Cost = Cost of Testing + Re-cook Charge + Penalty = Rs 6000 + 80 × Rs 75 + 20 × Rs 150 = Rs 15,000.
For x = 100, penalty = Rs 150 × 100 = Rs 15000

A2. (b)

Check for x = 300.
Total cost using Test 1:
Cost of testing = Rs 6 × 1000 = Rs 6000
Out of total defective 300 boxes, 80% will be detected. So, total detected boxes = 80% of 300 = 240 boxes. These 240 boxes will be reworked upon and remaining 60 boxes will go undetected. As per the rule, company will have to pay penalty on these 60 boxes.
Re- cook charges = Rs 75 × 240 = Rs 18000
Penalty = Rs 150 × 60 = Rs 9000
So, total cost using Test 1 = Cost of Testing + Re-cook Charge + Penalty
= Rs 6000 + Rs 18000 + Rs 9000 = Rs 33,000.
Total cost using Test 2:
Cost of testing = Rs 9 × 1000 = Rs 9000
Out of total defective 300 boxes, 100% will be detected. So, total detected boxes = 100% of 300 = 300 boxes. These 300 boxes will be reworked upon.
Re-cook charges = Rs 75 × 300 = Rs 22500
So, total cost using Test 2 = Cost of Testing + Re-cook Charge + Penalty
= Rs 9000 + Rs 22500 + Rs 0 = Rs 31,500.
It can be seen that Total cost of Test 1 < Total cost of Test 2. Hence test 2 should be used.
Hence option (b) is the answer.

A3. (a)

A4. (b)

Using solution to Q (1), Penalty for x = 100 is equal to Total Cost of Test 1 for x = 100.
Hence answer is – No Test or Test 1 if x = 100.

Pie Chart Comparison

Set 5

Direction for questions 1–4

Go through the data set given below and solve the questions based on it.

The following pie charts represent the percentage share of each of the five car manufacturing companies—A, B, C, D and E—for the two consecutive years—2008 and 2009.

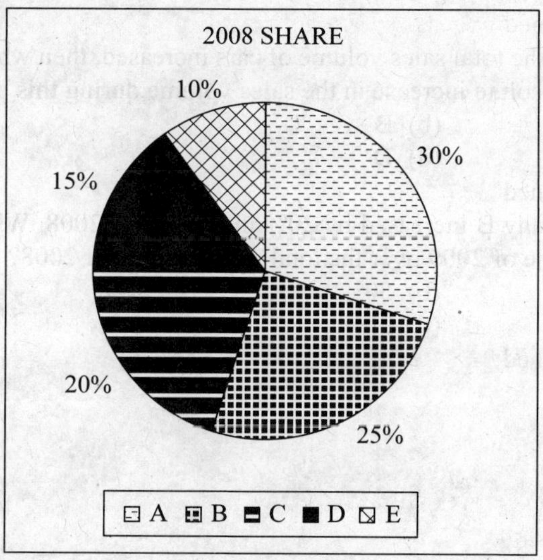

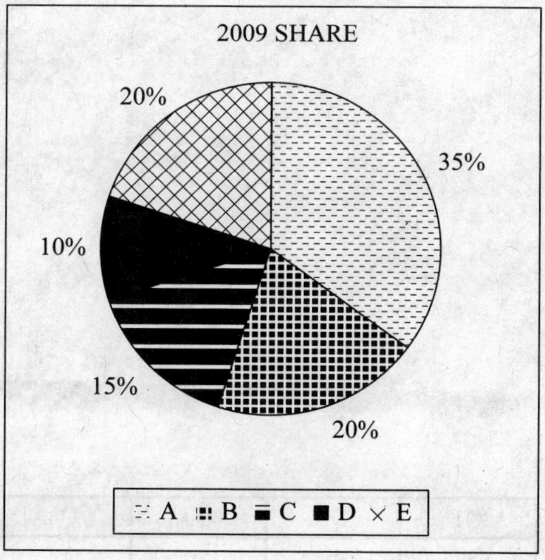

A = 35% and then clockwise for B, C, D and E.

Q1. For how many of the companies mentioned in the above pie charts, the sales volume has certainly increased in 2009 over 2008?

(a) 0 (b) 1
(c) 2 (d) 3
(e) Cannot be determined

Q2. From 2008 to 2009, if the sales volume of none of the manufacturing companies decreased, then what is the minimum percentage increase in the total sales volume of cars in 2009 over 2008?

(a) 25% (b) 50%
(c) 75% (d) 20%
(e) Cannot be determined

Q3. From 2008 to 2009, if the total sales volume of cars increased, then which of the following companies had the highest absolute increase in the sales volume during this period?

(a) A (b) B
(c) C (d) D
(e) Cannot be determined

Q4. Sales volume of company B increased by 20% in 2009 over 2008. What is the percentage change in the total sales volume of 2009 over the total sales volume of 2008?

(a) 25% (b) 50%
(c) 75% (d) 20%
(e) Cannot be determined

Rough Work

Answer Grid

Q1	Q2	Q3	Q4
e	b	e	b

STORYLINE

Pie charts give the percentage composition of the car market—it tells the market share of the different companies. Since total number of cars sold is not given, we cannot calculate the number of cars sold by each of the companies for any of the years. You may expect questions which contain data for a particular year of a company to be connected with the data of the other company.

Solutions to Q1–Q4

A1. Though in some cases, percentage market share has increased but unless we have the numbers of cars sold for a particular year, we cannot compare the sales data of any company of 2008 to the data of 2009. So the answer is option (e) cannot be determined.

A2. Understand this questions 1st – It can be seen that market share of some of these companies have gone down in 2009 over 2008. Despite this, sales volume of none of these companies has gone down. It is possible only if the total number of production has increased considerably so that even if market share is down in 2009, but because it is out of a 'larger pie', hence number of cars sold will be higher in 2009 than in 2008.
Market share of companies A and E have grown up.
Market share of companies B, C and D have gone down.
Maximum percentage decrease in market share occurs in the case of company D. Percentage decrease in the market share of D = $\frac{15-10}{15} \times 100 = 33.33\%$. So we should be working out with D.
Assume that the total car sales in the year 2008 – 100 units. So number of cars sold by D = 15.
Since the number of cars sold does not go down for any of the companies, including that of D, we assume that the number of cars sold in 2009 by D = 15. Assume that the total number of cars sold in 2009 = x
Hence 10% of $x = 15 \Rightarrow x = 150$ = Total number of cars sold in 2009.
Hence percentage increase in the total cars sold = $\frac{150-100}{100} \times 100 = 50\%$.
Had it been anything less than 50%, the total cars sold by D in 2009 would have been less than the number of cars sold by D in 2008.

A3. Market share of only two companies increased during this period—A and E. So answer should be one of these. Consider the following cases:

			Assume 2008 = 100 and 2009 = 120		Assume 2008 = 100 and 2009 = 1000	
	A	E	A	E	A	E
2008	30%	10%	30	10	30	10
2009	35%	20%	42	24	350	200
		Sales Volume	12	14	320	190
		Verdict	E > A		A > E	

Hence answer is option (e) cannot be determined.

A4. Assume that the total sales volume in 2008 = 100 units.
Market share of B in 2008 = 25%. Hence sales volume of B in 2008 = 25% of 100 = 25
Sales volume of B has increased by 20% in 2009 over 2008.
Hence sales volume of B in 2009 = 1.2 × 25 = 30
Assume that the total sales volume in 2009 = x.
So, 20% of x = 30 = Sales volume of B in 2009. So, x = 150.

Hence percentage increase in the total cars sold = $\frac{150-100}{15} \times 100 = 50\%$.
Hence option (b) is the answer.

BPL Population

Set 6

Direction for questions 1–5

Study the following chart carefully and answer the questions based on it.

Following table shows the percentage population of six states below poverty line and the proportion of male and female.

State	Percentage Total Population Below poverty Line	Proportion of Male and Female	
		Below Poverty Line	Above Poverty Line
		Male: Female	Male: Female
A	12	3:2	4:3
B	15	5:7	3:4
C	25	4:5	2:3
D	26	1:2	5:6
E	10	6:5	3:2
F	32	2:3	4:5

Everybody is either above the poverty line or below the poverty line.

Answer to some of the questions may be fractional. For the sake of calculation convenience, take integral values in such cases.

Q1. The total population of State A is 3200 thousand, then what is the **approximate** number of females above poverty line in state A (in thousands)?
(a) 1206 (b) 2112 (c) 1800 (d) 1950

Q2. If the total population of C and D together is 18000, then what is the total number of females below poverty line in these two states?
(a) 5000 (b) 5500 (c) 4800 (d) Cannot be determined

Q3. If the population of males below poverty line in state B is 500 then what is the total population of B?
(a) 14400 (b) 6000 (c) 8000 (d) 7600

Q4. If the population of males below poverty line in state A is 3000 and that in state E is 6000, then what is the ratio of the total population of state A and E?
(a) 3:4 (b) 4:5 (c) 21:25 (d) None of these

Q5. If in state E, population of females above poverty line is 19800 then what is the population of males below poverty line in that state?
(a) 5500 (b) 3000 (c) 2970 (d) None of these

Rough Work

Answer Grid

Q1	Q2	Q3	Q4	Q5
a	d	c	d	b

Solutions to Q1–Q5

A1. Total population above poverty line = 100% – 12% = 88%
So, total population above poverty line = 88% of 3200 = 2816
Ratio of Male: Female above poverty line = 4:3

Number of females above poverty line in state A = $\frac{3}{7} \times 2816 = 1206$
Hence option (a) is the answer.

A2. From the given information, we cannot find the population of states C and D individually; so we can't find the required value. Hence option (d) is the answer.

A3. Population of State B below poverty line = 15%

Ratio of males to females below poverty line = 5:7 ⇒ So males are $\frac{5}{12}$ of the total population below poverty line.

So, $\frac{5}{12}$ of 15% population of A = 500

So, total population of A = $500 \times \frac{12}{5} \times \frac{100}{5} = 8000$. Hence option (c) is the answer.

A4. Population of State A below poverty line = 12%

Ratio of males to females below poverty line = 3:2 ⇒ So males are $\frac{3}{5}$ of the total population below poverty line.

So, $\frac{3}{5}$ of 12% population of A = 3000

So, total population of A = $3000 \times \frac{5}{3} \times \frac{100}{12}$

[At this stage, we are not going to calculate the exact value of population of A. Because, we have to find out the ratio of population of A and E.]

Similarly, total population of E = $6000 \times \frac{11}{6} \times \frac{100}{10}$

Therefore, required ratio = $\frac{3000 \times 5 \times 100 \times 6}{6000 \times 3 \times 12 \times 11 \times 10} = \frac{25}{66}$

Hence option (d) is the answer.

A5. Population of State E below poverty line = 10%. Hence population of state E above poverty line = 90% of the total population.

Ratio of males to females above poverty line = 3:2 ⇒ So females are $\frac{2}{5}$ of the total population above poverty line.

So, $\frac{2}{5}$ of 90% population of E = 19800

So, total population of E = $19800 \times \frac{5}{2} \times \frac{100}{90} = 55000$

So, number of males below poverty line = 10% of total = 10% of 55000 = 5500

Ratio of males: females below poverty line = 6 : 5.

Hence males are $\frac{6}{11}$ of the total population below poverty line = $\frac{6}{11} \times 5500 = 3000$.

Hence option (b) is the answer.

Demand–Supply–Capacity Set 7

Direction for questions 1–5

Go through the chart given below and solve the questions based on it.

Following bar chart presents supply as a percentage of demand of a product and demand as a percentage of capacity of the industry to produce the same product.

All the values are multiples of 5.

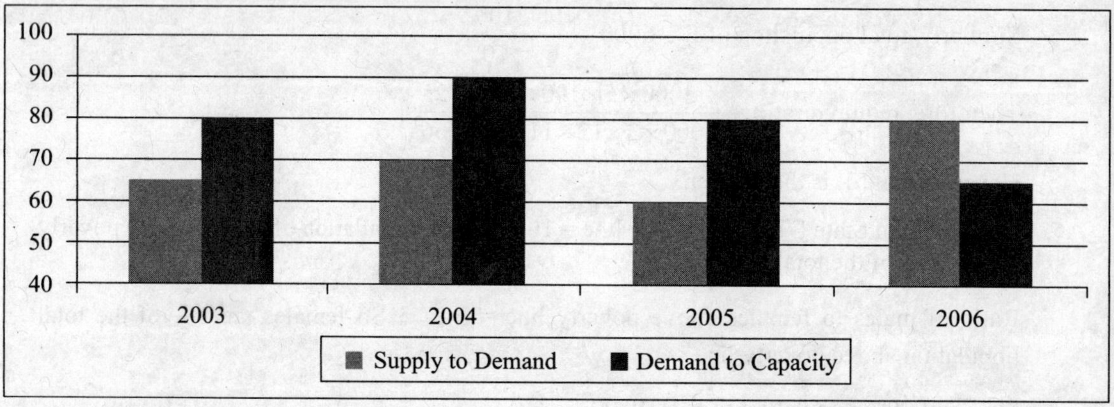

Q1. If product supply for the year 2005 is 6000 kg, then the total capacity of the product is (in kg):

(a) 12500 (b) 16500 (c) 10000 (d) None of these

Q2. If capacity of the industry to produce the product remains constant at 10000 kg, what is the approx. average supply (in kg) for the given period?

(a) 5100 (b) 5375 (c) 4400 (d) 6200

Q3. If the product supply remains unchanged for the given period then what does the trend of industry capacity signify?

(a) Industry capacity steadily increases over the given period.
(b) Industry capacity steadily decreases over the given period
(c) Industry capacity fluctuates over the given period.
(d) None of these

Q4. Which year shows the highest increase in the difference between the industry capacity and supply over the previous year, if demand for the product is 10000 kg for each of the given years?

(a) 2003 (b) 2004 (c) 2005 (d) 2006

Q5. If supply is constant for each of the years given above, then for which year is the capacity highest?

(a) 2003 (b) 2004 (c) 2005 (d) 2006

Rough Work

Answer Grid

Q1	Q2	Q3	Q4	Q5
a	b	c	c	c

STORYLINE

First bar mentions the ratio supply to demand i.e., supply: Demand. For example, as mentioned in the bar of 2003, 65% is the value for Supply to Demand. It means that supply is 65% of demand. In other words, it means that if demand is of 100 units, then supply is 65 units. Similarly, the second bar gives the value for Demand to Capacity. For example, as mentioned in the bar of 2003, 80% is the value for Demand to Capacity. It means that if capacity is of 100 units, then demand is 80 units.

It can also be observed that if we join the values in both the bars, then a relation between Supply and capacity can be established using ratio or percentage calculation.

Solutions to Q1–Q5

A1. In year 2005, percentage of supply to demand = 60%. So supply is 60% of the demand.
So, 60% of demand = 6000 kg ⇒ 100% of demand = Total Demand = 10000 kg.
In year 2005, percentage of demand to capacity = 80%. So demand is 80% of the capacity.
So, 80% of the capacity = 10000 kg ⇒ 100% of the capacity = Total capacity = $10000 \times \frac{100}{80}$ = 12500 kg.
Alternatively, in year 2005, 80% of capacity = Demand, and 60% of demand = Supply.

Assume that capacity = p, then demand = 80% of p = 0.8p.
Further, 60% of 0.8p = 0.48p = Supply

Using the information given in the question, 0.48p = 6000 ⇒ p = 6000 × $\frac{100}{48}$ = 12500.
Hence option (a) is the answer.

A2. In year 2003, Supply = 65% of demand. Further, Demand = 80% of Capacity ⇒

Supply = 65% of 80% of capacity = $\frac{65}{100} \times 80$ of capacity = 52% of Capacity

Similarly, in year 2004, Supply = 70% of 90% of Capacity = 63% of Capacity
In year 2005, Supply = 60% of 80% of Capacity = 48% of Capacity
In year 2006, Supply = 80% of 65% of Capacity = 52% of Capacity
Since capacity of the industry is constant every year, to find out the average supply per year, we can simply find out the average percentage supply and multiply it with 10000.

Average percentage supply per annum = $\frac{52\% + 63\% + 48\% + 52\%}{4} = \frac{215\%}{4} = 53.75\%$

Hence, Average supply per annum = 53.75% of 10000 = 5375. Hence option (b) is the answer.

A3. Using data from previous question:
Supply as a percentage of demand:

2003	2004	2005	2006
52%	63%	48%	52%

It is given that supply remains constant. Since there is no fixed pattern of supply as a percentage of demand (as given in above table), hence there is no fixed trend of capacity.

A4. Following data can be obtained:

	2003	2004	2005	2006
Supply	6500	7000	6000	8000
Capacity	12500	11111.11	12500	15384.62
Difference	6000	4111.11	6500	7384.6

It can be easily seen that maximum increase occurs in 2005. Hence option (c) is the answer.

A5. Given that supply is same for each of the years. Using the data for Q3, supply as a percentage of demand is minimum in 2005 (=48%). Hence total supply for the year 2005 will be maximum. Hence option (c) is the answer.

Funds Utilized

Set 8

Direction for questions 1–4

Go through the table and information given below and solve the questions based on it.

The table below gives the state-wise cumulative utilization of funds available under the Members of Parliament–Local Area Development Scheme (MP-LADS), by the Lok Sabha MPs, for the fiscal years 2005–06 and 2006–07.

States having	Fiscal Year 2005–06		Fiscal Year 2006–07	
MP-LADS	Amount Available	% Funds Utilized	Amount Available	% Funds Utilized
Andhra Pradesh	888	95	974	93
Bihar	812	70	880	78
Gujarat	542	95	586	90
Karnataka	578	93	628	89
Madhya Pradesh	611	75	667	82
Maharashtra	977	89	1080	85
Rajasthan	530	82	587	88
Uttar Pradesh	1700	78	1843	87
Total (8 States)	6638	83.65	7245	86.4

Q1. The percentage of funds utilized by Andhra Pradesh, Gujarat, Karnataka, and Maharashtra has changed in the following way from fiscal year 2005–06 to 2006–07.
(a) Increased by around 10% (b) Remained the same
(c) Decreased by around 10% (d) Decreased by around 5%

Q2. The ratio of the funds utilized by Bihar, as the percentage of the funds utilized by all the states in fiscal year 2005–06 to 2006–07 has
(a) Decreased by around 1% (b) Decreased by around 5%
(c) Increased by around 5% (d) Increased by around 1%

Q3. The ratio of the funds available to Uttar Pradesh against those available to all other states in fiscal years 2005–06 and 2006–07 has
(a) Decreased by around 1% (b) Decreased by around 5%
(c) Increased by around 5% (d) Increased by around 1%

Q4. Out of the given states, for how many states, funds used in 2006–07 are more than the funds used in previous year?
(a) 5 (b) 6 (c) 7 (d) 8

Rough Work

Answer Grid

Q1	Q1	Q3	Q4
d	c	a	d

Spam e-mails

Set 9

Directions for questions 1–3

Answer the questions on the basis of the information given below.

Spam that enters our electronic mailboxes can be classified under several spam heads. The following table shows the distribution (out of 100%) of such spam worldwide over time. The total number of spam emails received during December 2002 was larger than the number received in June 2003. The total number of spam emails received during September 2002 was larger than the number received in March 2003. The figures in the table represent the percentage of all spam emails received during that period, falling into those respective categories.

Category	Sep-02	Dec-02	Mar-03	Jun-03
Adult	38	33	19	17
Health	25	30	37	45
Financial	11	19	5	18
Internet	5	3	10	6
Products	3	7	10	11
Scams	5	6	11	2
Others	13	2	8	1

Q1. In which category was the percentage of spam e-mails increasing but at a decreasing rate?
(a) Financial (b) Scams (c) Products (d) None of the above

Q2. In the health category, the number of spam e-mails received in December 2002 as compared to June 2003 is
(a) was larger (b) was smaller (c) was equal (d) Cannot be determined

Q3. In the financial category, the number of spam emails received in September 2002 as compared to March 2003
(a) was larger (b) was smaller (c) was equal (d) Cannot be determined

Rough Work

Answer Grid

Q1	Q2	Q3
c	d	a

STORYLINE

Data provided in the table presents the percentage of categories of different types of spam emails for the given period. Since this does not provide us the absolute figures and total number of spam emails during these four month (as mentioned in the question) are different, we cannot compare the number of spam emails across categories by just looking at the numbers given in the table.

Still, Q.1 can be solved by just looking at the numbers as it does not require comparing the data across different categories.

First of all, let us tabulate the data given in the passage:

Number of spam e-mails in Dec 2002 > Number received in June 2003

Number of spam e-mails in Sep 2002 > Number received in March 2003

Solutions Q1–Q3

A1. There are only two categories where percentage of spam emails is increasing constantly—Health and Products. Since option (d) is None of these, we are required do some calculation that which of the two options – (c) Product or (d) None of these will be the answer. Had the 4th option been any of the other categories like Adult or Internet or Others, we would have marked products as the answer without doing any calculation.

In case of products, percentage of spam e-mails is increasing but at decreasing rate, from Sep 2002 to Dec 2002 products increased by $\frac{7-3}{3}\times 100\ \times 100 = 133.33\%$ and in March 2003 about $\frac{7-4}{7}\times 100 = \frac{3}{7}\times 100 = 42.84\%$ and in June 2003, $[\frac{11-10}{10}\times 100] = 10\%$.

Hence option (c) Products is the answer.

A2. Since percentage of spam in Dec 2002 is LOWER than June 2003, and the number of total e-mails received is higher, hence cannot be determined.

A3. In September 2002, percentage is higher as compared to March 2003, however the total number of e-mails received in Sept 2002 is lower than that in March 2003.

Let us see it this way:

Number of spam e-mails in Sep 2002 (assume = 100) > Number received in March 2003 (assume = 50)

So, we have to compare 11% of 100 and 5% of 50. It can be clearly seen that 11% of 100 is larger. If could have been seen even without assuming values also that 11% of a larger value is greater than 5% of a smaller value.

11% of x > 5% of y [where x > y]

Hence obviously larger. Hence option (a) is true.

It should also be mentioned here that, in above example, if y > x, then we cannot conclude anything ⇒ 11% of x and 5% of y [y > x] nothing can be said. Similar situation arises in Q2, and hence we cannot conclude which one—December 2002 as compared to June 2003 in health category is larger in Q2.

Six Departmental Stores Set 10

Direction for questions 1–5

Go through the data given below and solve the questions based on it.

The percentage change in sales of six departmental stores from 2003 to 2005 is shown in the following table:

Store	Percentage change	
	From 2003 to 2004	From 2004 to 2005
UCB	+20	–20
Van	–20	+9
Wrangler	+5	+12
X-tra	–7	–15
Yo	+17	–8
Zodiac	+21	+9

Q1. In 2005, which of the stores had greater sales than any of the other stores?
(a) Van (b) Wrangler (c) Zodiac (d) Cannot be determined

Q2. In store Yo, the sales for 2004 amounted to about what percentage the sales for 2005?
(a) 117 (b) 109 (c) 101 (d) 113

Q3. If in 2003, the sales in store Wrangler amounted to Rs x, the sales for that store in 2005 amounted to:
(a) $1.5x$ (b) $1.17x$ (c) $1.32x$ (d) $1.61x$

Q4 Which of the following has the maximum percentage increase in the sales in 2005 over 2003?
(a) Van (b) Wrangler (c) Zodiac (d) Cannot be determined

Q5. For how many of the stores given in the table can we definitely determine that if that particular store's sales in 2005 is more than the its sales in 2003?
(a) 0 (b) 1 (c) 2 (d) None of these

Rough Work

Answer Grid

Q1	Q2	Q3	Q4	Q5
d	b	b	c	d

STORYLINE

Table presents the growth percentage (or negative growth) for six companies. Data presents the way the business is moving ahead for these companies on a year to year basis. Sales data has been compared in 2004 over the sales data in 2003. Similarly, sales data in 2005 has been given with respect to the sales data given in 2004. Since all the data given are given in percentage terms (percentage increase or percentage decrease), we cannot find out the exact sales data for any year unless we are given the data for any particular year.

Solutions to Q1–Q5

A1. Question is asking for exact sales value. Data given in the table presents the percentage change over the last year. Since we do not know the initial sales figure for any of these companies, we cannot find out the sales figure of 2005 for any of these companies. (We cannot assume that sales value of companies in 2003 is same). Hence option (d) is the answer.

A2. Question is asking for a direct comparison between the sales value of 2005 over the sales value of 2004. Since nowhere the exact sales value is involved, we can solve this question. Since question is asking for a comparison between 2005 over 2004, we are not required to involve the value of 2003.
Let us assume that the sales value of 2004 = Rs 100. So, sales value of 2005 = 8% less than the sales value of 2004 = 8% less than 100 = 92.
Question is – 'Sales for 2004 amounted to about what percentage of the sales for 2005?'
Since comparison is made with respect to the sales for 2005, sales for 2005 will be the base and will be placed as the denominator.

Answer = $\frac{\text{sales for 2004}}{\text{sales for 2005}} \times 100 = \frac{100}{92} \times 100 = 1.08 \times 100 = 108\%$. Looking at the options, option (b) is the closest one. Hence option (b) is the answer.

Even without doing the actual calculation, we can see that $\frac{100}{92} \times 100$ is much more than 101, hence 101% cannot be the answer. Hence option (c) is eliminated. Similarly, it is less than 110% as 110% will be equal to $\frac{92+9.2}{92} = \frac{101.2}{92}$. Hence option (a) and option (d) are eliminated.
This leaves us with option (b) as the answer.

A3. Now this is one sitter:

There are two most commonly used methods to solve questions like these:

Method 1–Assuming the initial value

Assume that initial value = 100 units

$100 \xrightarrow{5\%\ increase} 105\ 100 \xrightarrow{12\%\ increase} 117.6$

So, if initial value = x, final value = 1.17x

Method 2–Using mechanism for successive percentage change

If there are two successive increases of a% and b%, then net successive % increase = $(a + b + \frac{ab}{100})$

Using this, net successive percentage increase = $5 + 12 + \frac{5 \times 12}{100}\% = 17.6\%$

If x is the initial value, then after an increase of 17.6%, final value will be 1.176x. Hence option (b) is the answer.

Note on Approximation:

Before doing any final calculation, we must look at the options. It can be clearly seen that options given in this question are not really close, so we can do the approximation instead of going through the actual calculation.

In this question, from 2003 to 2005, there are two successive increases of 5% and 17% ⇒ This is roughly equal to an increase of (12 + 5)% = 17% [Though we understand that the actual value will be slightly more than 17%]. Looking at the options, option (b) 1.17x is the best answer.

A4. This question can be answered since it asks only for highest percentage increase, and not the sales value (as in Q1). So, answer cannot be option (d) cannot be determined.

Now we are left out with the three companies as given in the first three options.

Using the approximation tips given in the previous question,

Option (a) Van = Approx (–11%)

Option (b) Wrangler = Approx. 17%

Option (c) Zodiac = Approx. 30%. Hence option (c) is the answer.

A5. We can calculate it for all the years. Hence option (d) is the answer.

European Football Cup **Set 11**

Direction for questions 1–5

Go through the data set given below and solve the questions based on it.

In the European football cup, five teams participated. Each team played with other team twice. Teams were awarded points on the basis of the matches won (W), lost (L) and drawn (D) by them.

The numbers of matches won, lost and drawn by the five teams are shown in the following table:

Team Name	Won (W)	Lost (L)	Drawn (D)
Sri Lanka	4	3	1
Myanmar	2	2	4
Laos	3	3	2
Fiji	P	Q	R
Estonia	6	0	2

Note *For every match won, the team gets 4 points, for every match drawn, the team gets 2 points while for every match lost, there are no points.*

Q1. What are the number of matches won by Fiji?
(a) 5 (b) 3 (c) 2 (d) 0

Q2. The number of matches drawn by Fiji are
(a) 2 (b) 3 (c) 1 (d) 0

Q3. The number of matches lost by Fiji are
(a) 5 (b) 3 (c) 2 (d) 7

Q4. What is the total number of points earned by Fiji?
(a) 8 (b) 14 (c) 2 (d) 6

Q5. Rank the teams in the order of decreasing points.
(a) Sri Lanka, Myanmar, Laos, Fiji, Estonia
(b) Estonia, Sri Lanka, Laos, Myanmar, Fiji
(c) Myanmar, Sri Lanka, Laos, Fiji, Estonia
(d) Fiji, Estonia, Sri Lanka, Myanmar, Laos

Rough Work

Answer Grid

Q1	Q2	Q3	Q4	Q5
d	c	d	c	b

STORYLINE

Generally, there are two methods in which tournaments happen—Round Robin League Tournament (also known as League Tournament) and Knockout Tournament (also known as Elimination Tournament). In this question, first method—Round Robin League tournament method is used.

In Round Robin League Tournament, each team plays with every other team. For example, if there are 5 teams namely A, B, C, D and E, total matches will be = AB, AC, AD, AE, BC, BD, BE, CD, CE, DE. (AB means a match between A and B and so on. There cannot be a match between team A and team A, hence AA does not make sense).

So, if there are 5 teams, total number of matches = 10

In general, if there are n teams, total number of matches $^nC_2$

Where $^nc_2 = \frac{n(n-1)}{2}$

is also equal to [1 + 2 + 3 + 4 + ...+ (n – 2) + (n – 1)].

For example, if there are six teams in a round robin league tournament, total number of matches = 5 + 4 + 3 + 2 + 1 = 15. This is also equal to $^6c_2 = \frac{6(6-1)}{2} = \frac{6\times 5}{2} = 15$.

In this question, each team plays with every other team twice. So total number of matches played by each team = 4 × 2 = 8 matches. Though, it is noteworthy here total number of matches will not be 8 matches/team × 5 teams = 40 matches (because there is 'double counting' of each match). Total number of matches = = 20 matches

Solutions to Q1–Q5

A1. Total number of matches won = Total number of matches lost. Hence, 15 + P = 8 + Q

Total number of matches played:

Method 1–Total number of matches = $\frac{8\times 5}{2}$ = 20 matches

Method 2–4 × 2 + 3 × 2 + 2 × 2 + 1 × 2 = 20

Total number of matches drawn = (9 + R)/2.

Total number of matches played = matches lost + matches drawn.

Hence, 2Q + R =15. where P and Q are non-negative integers. This equation is satisfied when Q = 7 & R = 1. Hence, Ans (4)

A2. Ans (c)

A3. Ans (d)

A4. Ans (c)

A5. The points of Sri Lanka, Myanmar, Laos, Estonia, Fiji are 18, 16, 16, 28, 2 respectively. Ans (b)

Transportation Problem Set 12

Direction for questions 1–4

Go through the data given below and solve the questions based on it.

Table 1 give the transportation cost (in Rs) of one product from six factories F1, F2, F3, F4, F5 and F6 to five warehouses W1, W2, W3, W4 and W5.

	W1	W2	W3	W4	W5
F1	60	25	45	40	80
F2	85	40	45	60	75
F3	80	20	45	50	90
F4	90	40	55	25	70
F5	95	20	40	45	80
F6	85	35	50	50	75

Table 2 give the transportation cost (in Rs) of the same product from five warehouses W1, W2, W3, W4 and W5 to six retailers R1, R2, R3, R4, R5 and R6.

	R1	R2	R3	R4	R5	R6
W1	40	20	40	30	85	25
W2	50	50	45	60	70	30
W3	20	10	50	50	80	35
W4	40	30	45	40	90	40
W5	50	30	50	40	70	30

Total transportation cost is the summation of the transportation cost from Factory to Warehouse and Warehouse to Retailer.

Q1. What is the minimum transportation cost incurred in transferring the product from any factory to any retailer?

(a) Rs 45 (b) Rs 50 (c) Rs 55 (d) Rs 60

Q2. What is the maximum transportation cost incurred in transferring the product from any factory to any retailer?

(a) Rs 175 (b) Rs 180 (c) Rs 170 (d) None of these

Q3. What is the minimum transportation cost at which the product can be supplied to R4 from any factory?

(a) Rs 45 (b) Rs 50 (c) Rs 65 (d) Rs 70

Q4 What is the maximum transportation cost at which the product can be supplied to R3 from any factory?

(a) Rs 145 (b) Rs 150 (c) Rs 140 (d) None of these

Answer Grid

Q1	Q2	Q3	Q4
b	b	c	c

STORYLINE

A product is manufactured in any one of the factories and transported to any one of the warehouse. The cost incurred in transporting from Factory to Warehouse is given in the first table. For example, cost of transportation of one unit of product from Factory 1 to Warehouse 4 is Rs 40 (Grey cell in the table given below

	W1	W2	W3	W4	W5
F1	60	25	45	40	80
F2	85	40	45	60	75
F3	80	20	45	50	90
F4	90	40	55	25	70
F5	95	20	40	45	80
F6	85	35	50	50	75

Similarly, the transportation cost from Factory 3 to Warehouse 5 is Rs 90 (Grey cell in the table given below):

	W1	W2	W3	W4	W5
F1	60	25	45	40	80
F2	85	40	45	60	75
F3	80	20	45	50	90
F4	90	40	55	25	70
F5	95	20	40	45	80
F6	85	35	50	50	75

Similarly, second table gives the transportation of one product from any warehouse to any retailer.

For example, cost of transportation of One product from Warehouse 5 to Retailer 2 is Rs 30 (Grey Cell in the table given below):

	R1	R2	R3	R4	R5
W1	40	20	40	30	85
W2	50	50	45	60	70
W3	20	10	50	50	80
W4	40	30	45	40	90
W5	50	30	50	40	70

So, if we transport one product from F3 (Factory 3) to W5 (Warehouse 5) to R2 (Retailer 2), total cost of transportation = Cost of transportation from F3 to W5 + Cost of transportation from W5 to R2 = Rs 90 + Rs 30 = Rs 120.

All the questions in the set are purely observational. Users are required to spot the numbers correctly across the grid.

Solutions to Q1–Q4

A1. Lowest will occur for F5 ⇒ W3 ⇒ R2
Hence option (b) is the answer.

A2. Maximum will occur for F5 ⇒ W1 ⇒ R5
Hence option (b) is the answer.

A3. Minimum will occur for F4 ⇒ W4 ⇒ R4
Hence option (c) is the answer.

A4. Maximum will occur for F3 ⇒ W5 ⇒ R3
Hence option (c) is the answer.

Distribution of Sales Set 13

Direction for questions 1–3

Go through the data set given below and solve the questions based on it.

The sales of a company in 2007–08 were Rs 1200 crore. Following charts represent the distribution of sales for various regions, product and distribution methods-wise:

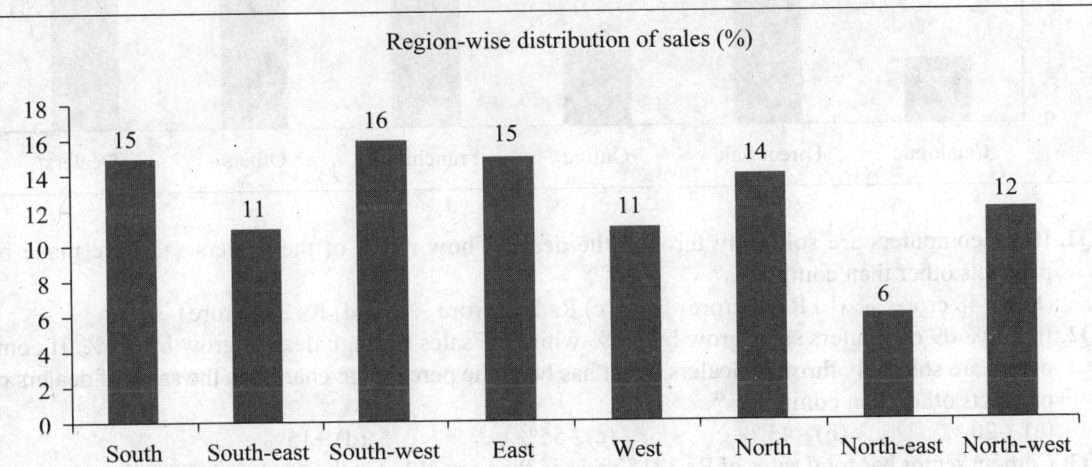

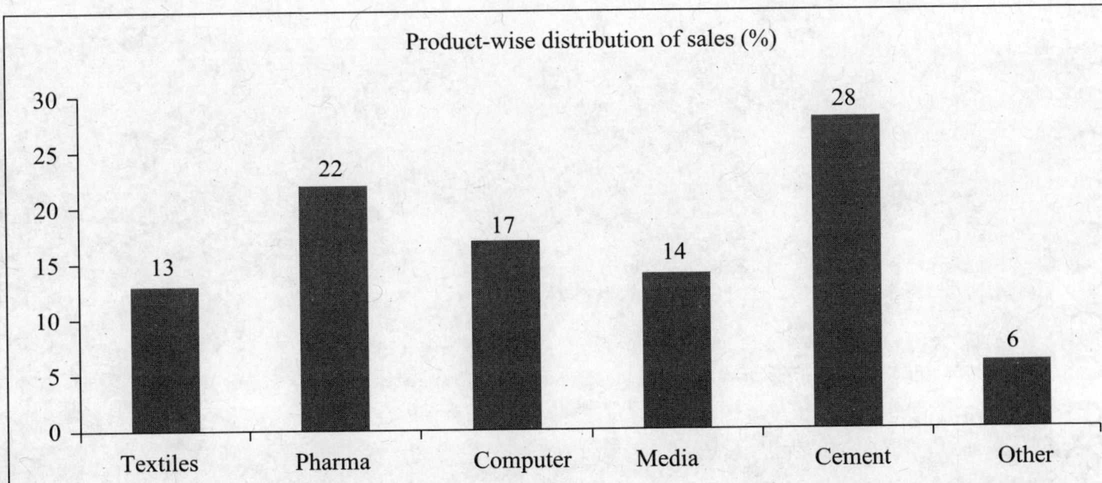

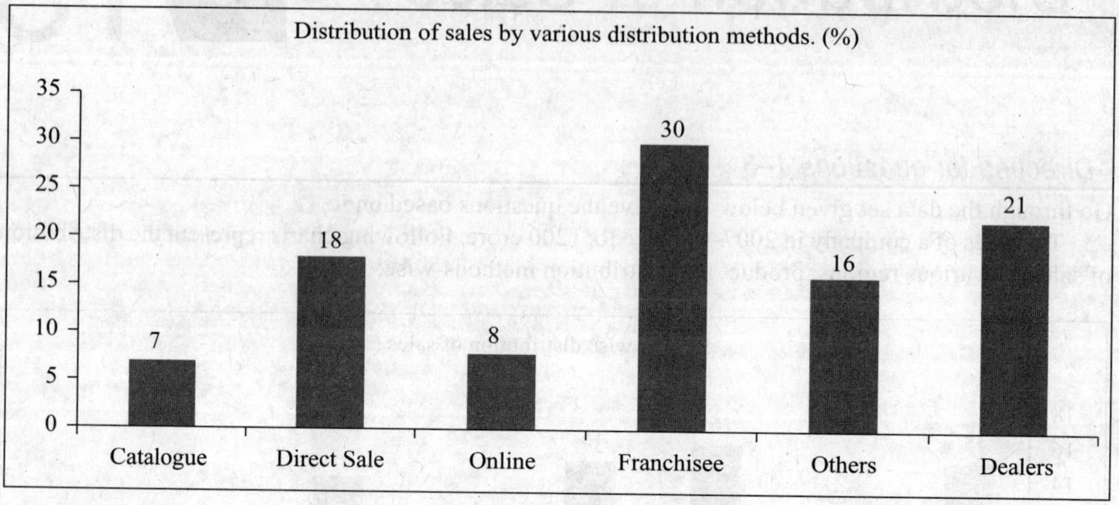

Q1. If the computers are sold only through the dealers, how much of the dealers sales are made by products other than computers?
(a) Rs 48 crore (b) Rs 43crore (c) Rs 252 crore (d) Rs 204 crore

Q2. In 2008–09 computers sales grow by 50 % while the sales through dealers grow by 30 %. If computers are sold only through dealers, what has been the percentage change in the sales of dealers of products other than computers?
(a) –20 % (b) –45 % (c) –55% (d) +15%

Q3. Cement sector has total sales of Rs 3215 crore in the year. What is the market share of the company?
(a) 9.55% (b) 10.45 % (c) 8.71 % (d) 11.55%

Answer Grid

Q1	Q2	Q3
a	c	b

STORYLINE

Three bar charts given present the same data in three different ways
(a) Region-wise – How much sales revenue is coming from which areas?
(b) Product-wise – How much percentage of total sales is coming through different products?
(c) Distribution channel-wise – How much percentage of total is coming through different distribution channel-wise?

Solutions to Q1–Q3

A1. All the computers are sold through dealers only.
Total sales volume of computers = 17% of Rs 1200 crores = 204 crores.
Question is asking for Dealers sale other than computer.
Total sales revenue through dealers = 21% of Rs 1200 crores = Rs 252 crores
Total dealers sale other than computer = Rs 252 crores – Rs 204 crores = Rs 48 crores.
Alternatively, it can be seen that total computers' sale has to be a subset of total dealers' sale.
Now, total dealers sale = 21% and total sales through computer = 17%.
Hence total sales by dealers of products other than computers = (21% – 17%) of Rs1200 crores = 4% of Rs 1200 crores = Rs 48 crores.
Hence option (a) is the answer.

A2. Computer's sale in 2007–08 = 17% of Rs 1200 crores = 204 crores.
Growth in 2008–09 = 50% over the sales value of previous year
So, computer sales in 2008–09 = 204 + 50% of 204 = Rs 306 crores
Total sales revenue through dealers in 2007–08 = 21% of Rs 1200 crores = Rs 252 crores
Growth in 2008–09 = 30% over the sales value of previous year
So, sales through dealers in 2008–09 = Rs 252 + 30% of Rs 252 = Rs 327.6
Total dealers sale other than computer in 2008–09 = Rs 327.6 crores – Rs 306 crores = Rs 21.6 crores.
Total dealers sale other than computer in 2007–08 = Rs 252 crores – Rs 204 crores = Rs 48 crores.[Using data from A1]
It can be seen that there is a drop from Rs 48 crores (in 2007–08) to Rs 21.6 crores (in 2008–09). Hence option (d) is ruled out.

Hence percentage increase = $\frac{21.6-48}{48}\times 100 = \frac{-26.4}{48}\times 100\%$

A quick glance at the options reveals that options are not close. Hence we can manage the answer with approximation.
50% of denominator 48 = 24. Numerator is more than this. Hence answer should be more than 50%. Hence option (c) is the answer.

A3. Total cement sales by this company = 28% of Rs 1200 crores = Rs 336 crores
Total cement sales = Rs 3215 crores

Hence market share of this company in cement sales = $\frac{336}{3215} \times 100$

10% of 3215 = Rs 321.5 crores. It can be easily seen that the value obtained will be greater than 10%. Hence option (a) and (c) are ruled out.
Now 336 = 321.5 + approx 15 = 10% + less than 1% < 11%
There is only one option less than 11%. Hence option (b) is the answer.

Estimated and Actual Profit

Set 14

Direction for questions 1–3

Go through the table given below and solve the questions based on it.

Table given below gives the Estimated and Actual Sales Revenue and Cost Figures for three years. Answer the questions based on it.

Sales Revenue	2007	2008	2009	Cost	2007	2008	2009
Estimated	1281	1452	1228	Estimated	986	1051	1024
Actual	1326	1286	1452	Actual	988	1076	1068

Estimated Profit = Estimated Revenue – Estimated Cost

Actual Profit = Actual Revenue – Actual Cost

Q1. For how many years, actual profit is more than estimated profit?

(a) 0 (b) 1 (c) 2 (d) 3

Q2. A measure of forecasting done by marketing team is the ratio of Estimated Sales to Actual Sales. Lower is this value, better is the forecasting. Which years' forecasting is the best?

(a) 2007 (b) 2008 (c) 2009 (d) Cannot be determined

Q3. Company plans to use a different mechanism for forecasting of its maximum possible profit = Max (Estimate, Actual (Sales Revenue)) – Min (Estimate, Actual(Cost)). What is the maximum possible profit derived this way?

(a) 340 (b) 401 (c) 428 (d) None of these

Rough Work

Answer Grid

Q1	Q2	Q3
c	c	c

STORYLINE

Question provides the Revenue and Cost Data in two ways: (a) Year wise, (b) Estimated and Actual with the definition of Estimated Profit and Actual Profit.

Solutions to Q1–Q3

A1. Following is the profit figure:

	2007	2008	2009
Estimated Profit	295	401	204
Actual Profit	338	210	384

Hence option (c) is the answer.

A2. Following is the ratio:

2007	2008	2009
0.966063	1.129082	0.84573

Hence option (c) is the answer.

A3. Max (Estimate, Actual (Sales Revenue))

2007	2008	2009
1326	1452	1452

Min (Estimate, Actual(Cost))

2007	2008	2009
986	1051	1024

Maximum possible profit = 428 crores (in 2009)
Hence option (c) is the answer.

Box Office Collections Set 15

Direction for questions 1–4

Go through the data set given below and solve the questions based on it.

Table given below presents the collections/business done by the top 12 movies in India. Column 3 gives the year in which movie was released, column 4 given the gross collection by the movie over the period of time since its release, column 5 gives the distributor's share out of the gross collection.

Answer the questions based on the data given in the table below.

All India (Figures in Rs)					
Rank	Movie Name	Release Year	Gross Collection (Rs Crores)	Distributor Share (Rs Crore)	Verdict
1	*Three Idlots*	2009	269	**99**	All Time Blockbuster
2	*Dabangg*	2010	187	**77**	All Time Blockbuster
3	*Ghajini*	2008	155	**60**	All Time Blockbuster
4	*Golmaal 3*	2010	144	**58**	Blockbuster
5	*Gadar Ek Prem Katha*	2001	130	**57**	All Time Blockbuster
6	*Hum Aapke Hain Kaun*	1994	123	**53**	All Time Blockbuster
7	*Raajneeti*	2010	124	**49**	Blockbuster
8	*Dhoom 2*	2006	111	**46**	Blockbuster
9	*Dilwale Dulhania Le Jayenge*	1995	106	**46**	All Time Blockbuster
10	*Rab Ne Bana Di Jodi*	2008	117	**44**	Blockbuster
11	*Om Shanti Om*	2007	110	**40**	Blockbuster
12	*Krissh*	2006	97	**40**	Blockbuster

*This is real data sourced from boxofficeindia.com.

Q1. For how many movies, distributor's share is more than 40% of the gross collection?

(a) 5 (b) 6 (c) 7 (d) 8

Q2. How many movies released in a non-leap year have been declared a blockbuster?

(a) 5 (b) 6 (c) 7 (d) 8

Q3. Since value of money across the years is not constant, we use Money Value Parity Unit (MVP) to compare the value of money across the year. Assuming 1990 as the base year with MVP = 100, all the years have been assigned a MVP. For example, MVP of 2000 = 58. It means that:

Value of Rs 100 of the year 2000 = Value of Rs 58 of 1990. So revenue of Rs 100 collected in the year 2000 is same as revenue of Rs 58 collected in the year 1990. Higher the collection, more successful is the movie.

Consider the data given below regarding the MVP of different years:

1990	100
1995	80
2001	60
2010	40

Assume that the all the collection occur in the same year as that of release year. Converting the collections on the base year 1990, arrange the following movies in descending order of success: *Dabangg, Golmaal 3, Gadar Ek Prem Katha, Dilwale Dulhania Le Jayenge*

(a) *Dabangg, Golmaal 3, Gadar Ek Prem Katha, Dilwale Dulhania Le Jayenge*
(b) *Dabangg, Golmaal 3, Dilwale Dulhania Le Jayenge, Gadar Ek Prem Katha*
(c) *Dilwale Dulhania Le Jayenge, Gadar Ek Prem Katha, Dabangg, Golmaal 3*
(d) *Dilwale Dulhania Le Jayenge, Dabangg, Gadar Ek Prem Katha, Golmaal 3*

Q4. How many All Time Blockbuster movies have surpassed gross collection of Rs 100?

(a) 5 (b) 6 (c) 4 (d) 7

Note from Author: *In this set, Q2 and Q4 are simple observation based questions, Q1 is calculation based question and Q3 is interpretation and calculation based question. I would strongly recommend that before you move on to see the solution, ensure that you have attempted all the questions to the best of your capability.*

Rough Work

Answer Grid

Q1	Q2	Q3	Q4
c	a	c	b

Solutions to Q1–Q4

A1. At first, it might appear that this question is a real calculation nightmare. Though for most of the movies, we can get away without doing actual calculation and get a Yes/No answer (if more than 40%) by just observing the data—through approximation.
Start calculating 10% of the gross collection and see if 4 times of this value (40%) is more than distributor's share.
For example, first approximate the gross collection data for *Three Idiots* = Rs 270 crores. 40% of this collection data = 40% of Rs 270 crores = Rs 108 crores. In the table, distributor's share is given as Rs 99 crores, which is way below Rs 108 crores. Hence, distributor's share is LESS than 40% of the gross collection.
Similarly, for *Dabangg*, approximate value of Total collection can be taken as Rs 190 crores and 40% of this = Rs 76 crores. In the table, distributor's share is given as Rs 77 crores, which is more than Rs 76 crores. Since distributor's share is more than 40% of a larger approximated value (Rs 190 crores), it will be more for a smaller real value (=Rs 187 crores). Hence, distributor's share is MORE than 40% of the gross collection.
Doing a similar analysis gives the following table:

Rank	Film	Year	% Share
1	*Three Idiots*	2009	36.8
2	*Dabangg*	2010	41.18
3	*Ghajini*	2008	38.71
4	*Golmaal 3*	2010	40.28
5	*Gadar Ek Prem Katha*	2001	43.85
6	*Hum Aapke Hain Kaun*	1994	43.09
7	*Raajneeti*	2010	39.52
8	*Dhoom 2*	2006	41.44
9	*Dilwale Dulhania Le Jayenge*	1995	43.4
10	*Rab Ne Bana Di Jodi*	2008	37.61
11	*Om Shanti Om*	2007	36.36
12	*Krissh*	2006	41.24

Grey cells show the movies that have distributor's share more than 40% of total collection. Hence there are 7 movies. Hence option (c) is the answer.

A2. *Golmaal 3, Raajneeti, Dhoom 2, Om Shanti Om, Krissh* are the five movies. Hence option (a) is the answer.

A3. Value of money has depreciated over the period (gone down). So if collection is Rs 100 in 2010 will be equal to collection of Rs 40 in 1990. So a collection of Rs 187 crores in 2010 would be equal to $\frac{40 \times 187}{100}$ = Rs 74.8 crores of 1990.

Movie Name	Release Year	MVP	Gross Collection	Value (On Base Year 1990)
Dabangg	2010	40	187	74.8
Golmaal 3	2010	40	144	57.6
Gadar Ek Prem Katha	2001	60	130	78
Dilwale Dulhania Le Jayenge (DDLJ)	1995	80	106	84.8

Hence the order is—*DDLJ* > *Gadar Ek Prem Katha* > *Dabangg* > *Golmaal 3*. Hence option (c) is the answer.

A4. Answer is 6 movies. Hence option (b) is the answer.

Six by Six

Set 16

Direction for questions 1–3

Go through the data set given below and solve the questions based on it.

Following table gives the marks obtained by 6 students A, B, C, D, E and F in 6 subjects—History, Science, Geography, English, Hindi and Math. Marks in bracket show total marks in that subject. For example—History (150) shows that total marks in History = 150.

Marks obtained by 6 students in 6 subjects

	History (150)	Science (200)	Geography (150)	English (200)	Hindi (100)	Maths (200)
A	75	110	90	140	75	170
B	105	130	75	130	80	140
C	95	105	80	150	90	160
D	85	115	95	125	65	135
E	115	135	110	145	70	125
F	120	160	96	110	55	145

Q1. How many students have got more than 50% marks in each of the subjects and more than 65% marks in total?

(a) 0 (b) 1 (c) 2 (d) 3

Q2. What is the difference between the percentage marks obtained by the student B in History and percentage marks obtained by C in Hindi?

(a) 25 (b) 20 (c) 35 (d) 30

Q3. For C, what is the difference between the percentage of marks in English and the percentage of marks obtained in total?

(a) 82 (b) $38\frac{1}{3}$ (c) 7 (d) 14

Rough Work

Answer Grid

Q1	Q2	Q3
d	b	c

STORYLINE

Data set simply presents the marks obtained by six students in six subjects. You are required to observe that full marks are *not* the same for all the subjects. You may expect questions based upon percentage marks obtained by one/some students in subjects.

Solutions to Q1–Q3

A1 C, E and F are the students. [We have to consider more than 50%]. Hence option (d) is the answer.

A2. $\frac{680}{1000} \times 100 = 68\%$

$C^{®} = 90\%$

Required difference = 90 – 70 = 20. Hence option (b) is the answer.

A3. Percentage of marks of English $\frac{150}{200} \times 100 = 75\%$

Average percentage of marks $\frac{680}{1000} \times 100 = 68\%$

Therefore required percentage = 75% – 68% = 7%

Hence option (c) is the answer.

Round Robin League Soccer Competition

Set 17

Direction for questions 1–4

Read the passage below and solve the questions based on it.

Following table gives the result of due North Soccer Competition. Matches were played round robin (each team playing against another once) and the teams participated were Bihar, Punjab and Haryana. The table has some data missing:

	Match Played	Won	Drawn	Lost	Goals for	Goals Against
Bihar	2		1		2	4
Punjab	2				3	7
Haryana	2					1

Q1. What is the total number of matches played in the competition?
(a) 3 (b) 6 (c) 9 (d) None of these

Q2. What did the score card read in favor of Punjab in Punjab-Haryana match?
(a) 2–5 (b) 1–5 (c) 1–4 (d) 1–6

Q3. How many matches did Punjab lose?
(a) 0 (b) 1 (c) 2 (d) 0 or 1

Q4. How many goals did Haryana score?
(a) 5 (b) 1 (c) 3 (d) 7

Rough Work

Answer Grid

Q1	Q2	Q3	Q4
a	b	b	d

STORYLINE

In this set of questions, we are first required to calculate total number of matches. Let us understand the scheme of matches in a Round Robin League Tournament.

Method 1–In a Round Robin League Tournament, each team plays against other team once. So if there are five teams A, B, C, D and E (let us assume), total number of matches are going to be the following:

AB, AC, AD, AE
BC, BD, BE
CD, CE,
DE

Hence total number of matches = 4 + 3 + 2 + 1 = 10.

In general, if there are n teams, then total number of matches = $(n - 1) + (n - 2) + \ldots. + 3 + 2 + 1 = \frac{n(n-1)}{2}$

Method 2–Each team plays one match with every other team. So, number of matches played by each team = 4. So total number of matches = 5 teams × 4 matches = 20 matches. But each of these matches has been counted twice—for example match AB and BA both have been counted.

So total number of unique matches = $\frac{20}{2}$ = 10 matches.

Method 3–Using Permutation and Combination

There is a match for every set of two players. So, if there are n players, total number of matches = nC2.

In this question, there are three teams—Bihar, Punjab and Haryana.

Hence total number of matches = 2 + 1 (Using method 1)

This is also equal to $\frac{n(n-1)}{2} = \frac{3(3-1)}{2}$ = 3 matches.

Hence answer to Q1.Total number of matches played in the competition = 3 matches. Hence option (a) is the answer.

Secondly,

Total goals for by all the teams = Total goals against all the teams

Now total goals against = 4 + 7 + 1 = 12 goals

This gives us the number of goals for Haryana = 12 – (2 + 3) = 7

	Match Played	Won	Drawn	Lost	Goals for	Goals Against
Bihar	2		1		2	4
Punjab	2				3	7
Haryana	2				7	1

Now, Bihar has drawn one of its matches, so it has either won or lost the other match. Since it has drawn one of its two matches, possible score for Bihar is:

	Match 1 (Draw)	Match 2 (Lost)	Hypothesis
Possibility 1	0–0	2–4	Hypothesis 1
Possibility 2	1–1	1–3	Hypothesis 2
Possibility 3	2–2	0–2	Hypothesis 3

2–2 = Goals for 2 and goals against 2.
It can be seen that Bihar is losing its second match.
Testing Hypothesis 1:
Hereafter, we will have to go ahead with hypothesis 1. Using the conditions given, then, we will verify if the assumptions that we have assumed are correct or not.
Assume that Bihar has its match drawn with Punjab and lost its match against Haryana.

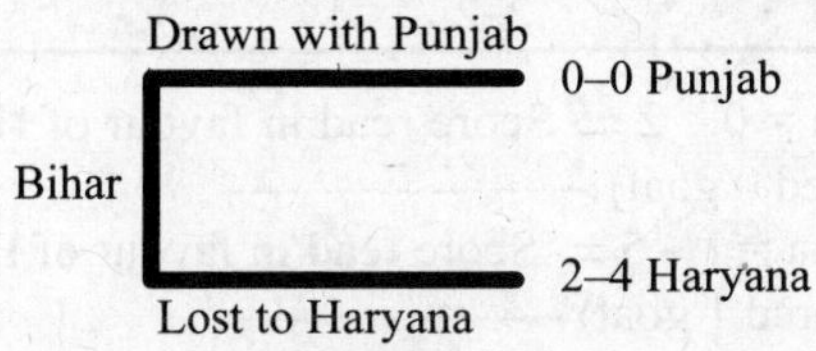

Since Punjab and Haryana both have played one match with Bihar at this stage, total number of matches occurred so far = 2. And next match will be between Punjab and Haryana.

Using the score card of Punjab – Goals for 3 and Goals against 7.

Since one of its matches was drawn with Bihar (score 0–0), hence score of 2nd match against Haryana would be = 3–7

	Match 1 (Drawn with Bihar)	Match 2 (Lost to Haryana)
Punjab	0–0	3–7

Now finally, we will have to evaluate if the goals for/against obtained in two matches against Haryana (one against Bihar and another against Punjab) support the figure given in the original table given in the question.
Bihar's score against Haryana = 2 – 4 ⇒ Score read in favour of Haryana = 4 – 2 (i.e., Haryana scored 4 goals and Bihar scored 2 goals) --------------------(1)

Punjab's score against Haryana = 3 – 7 ⇒ Score read in favour of Haryana = 7 -3 (i.e., Haryana scored 7 goals and Punjab scored 3 goals) ------------------(2)

Using the information given above in (1) and (2):

Haryana scored a total of 4 + 7 = 11 goals against these two teams. Looking at the information given in the table, Haryana has scored only 7 goals against both the teams taken together.

So, our hypothesis that Bihar drawn its match against Punjab with 0 – 0 is wrong.

Testing Hypothesis 2:

Similarly, we will test our 2nd hypothesis that Bihar drawn its match against Punjab with 1–1.

Testing Hypothesis 3:

Now checking the hypothesis that Bihar drawn its match against Punjab with score 2–2.

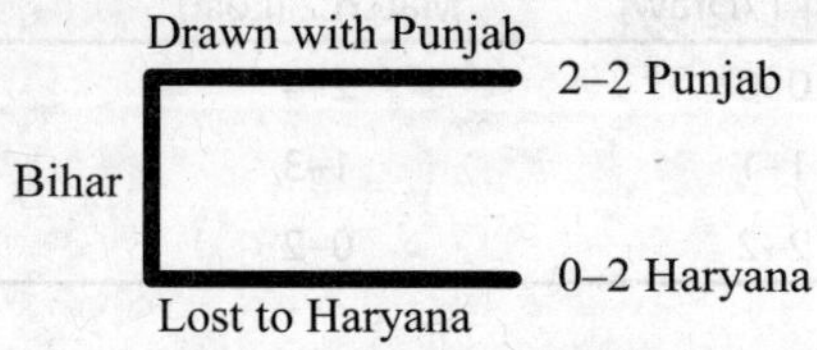

Using the score card of Punjab – Goals for 3 and Goals against 7.

Since one of its matches was drawn with Bihar (score 0–0), hence score of 2nd match against Haryana would be = 3–7

	Match 1 (Drawn with Bihar)	Match 2 (Lost to Haryana)
Punjab	2–2	1–5

Bihar's score against Haryana = 0 – 2 ⇒ Score read in favour of Haryana = 2 – 0 (i.e., Haryana scored 2 goals and Bihar scored 0 goal) --------------------- (1)

Punjab's score against Haryana = 1 – 5 ⇒ Score read in favour of Haryana = 5 – 1 (i.e., Haryana scored 5 goals and Punjab scored 1 goal) ------------------- (2)

Using the information given above in (1) and (2):

Haryana scored a total of 5 + 2 = 7 goals against these two teams. Looking at the information given in the table, Haryana has scored 7 goals against both the teams taken together. At the same time, Punjab and Bihar taken together have scored 0 + 1 = 1 goal against Haryana. This also suits the tally given in the table (in original question).

So hypothesis 3 is correct.

Hence the final table is like:

	Match Played	Won	Drawn	Lost	Goals for	Goals Against
Bihar	2	***0***	1	***1***	2	4
Punjab	2	***0***	***1***	***1***	3	7
Haryana	2	***2***	***0***	***0***	***7***	1

[New findings are in bold and italics].

Now all the questions from Q2 to Q4 can be answered.

Movie Buffs

Set 18

Direction for questions 1–4

Go through the data set given below and solve the questions based on it.

A survey of movie goers from five cities A, B, C, D and E is summarized below. The first column gives the percentage of viewers in each city who watch less than one movie a week. The second column gives the total number of viewers who view one or more than one movies per week.

City	I	II
A	60	2400
B	20	3000
C	85	2400
D	55	2700
F	75	8000

Assume people watch only integral number of movies.

Q1. How many viewers in the city C watch less than one movie a week?
(a) 2040 (b) 13600 (c) 16000 (d) 3600

Q2. Which city has the highest number of viewers who watch less than one movie a week?
(a) City E (b) City D (c) City B (d) City C

Q3. City with second highest number of movie watchers is
(a) City E (b) City D (c) City B (d) City C

Q4. The total number of people in the five cities who watch less than one movie per week is
(a) 46255 (b) 36000 (c) 18500 (d) 44155

Rough Work

Answer Grid

Q1	Q2	Q3	Q4
b	a	c	d

STORYLINE

Table provides name of five cities and data related to moviegoers in those cities. Segregation of data is done on the basis of number of people who watch less than one movie (i.e., No movie at all) per week and one movie or more per week. In other words, there are two categories of people—people who watch no movie and who watch one or more movie per week.

Column 2 gives the percentage of people who watch no movie (less than one movie). For example, 60% people in city A watch no movie. Hence 100% – 60% = 40% people watch one or more movie in the city.

Solutions to Q1–Q4

A1. In city C, percentage of people who watch less than one movie a week = 100% – 85% = 15%
Hence 15% = 2400.
Question is asking us to calculate number of people who watch less than a movie = Value in the 2nd column = 85%
Since 15% = 2400 ⇒ 5% = 800 ⇒ 85% = 17 × 800 = 13600.
Hence option (b) is the answer.

A2. A quick analysis of the data presents the following:

City	% who watch one or more	Number who watch one or more
A	40%	2400
B	80%	3000
C	15%	2400
D	45%	2700
E	25%	8000

Now, for city A, if 40% = 2400, then 60% = 1.5 times of 2400 [Using unitary method, $\left[\frac{60}{40}=1.5\right]=3600$.

Cities A, B and D can be outright eliminated on the ground as mentioned above.
We are required to choose between City C and city E.
In Q1, we have already calculated for city C. Number of viewers in city C who watch more than or equal to one movie = 13600.

For city E, since 25% = 8000, hence 75% = Number of people who watch less than a movie = 3 times of 8000 = 24000 [Using unitary method, $\left[\frac{75}{25} = 3\right]$]

Hence city E has the highest number of viewers. Hence option (a) is the answer.

A3. Now this is plain and simple. City E has the highest number of movie goers followed by city B. Hence option (c) is the answer.

A4. First, we explore the opportunity that can we solve this question through approximation? Since options (a) and (d) are quite close, and option (c) ruled out [answer to Q1 = 13600], for the sake of accuracy, it is advisable that we do this question by actual calculation.
Population watching less than one movie/week:
In city A = 3600, for city B = 750, for city C = 13600, for city D = 2205, for city E = 24000
So, total population watching less than a movie per week = 3600 + 750 + 13600 + 2205 + 24000 = 44155. Hence option (d) is the answer.

Double the Flowers

Set 19

Direction for question 1–3

Go through the information given below and solve the questions based on it.

A young girl Roopa leaves home with x flowers, goes to the bank of a nearby river. On the bank of the river, there are four places of worship, standing in a row. She dips all the x flowers into the river. The number of flowers doubles. Then she enters the first place of worship, offers y flowers to the deity. She dips the remaining flowers into the river, and again the number of flowers doubles. She goes to the second place of worship, offers y flowers to the deity. She dips the remaining flowers into the river, and again the number of flower doubles. She goes to the third place of worship, offers y flowers to the deity. She dips the remaining flowers into the river, and again the number of flowers doubles. She goes to the fourth place of worship, offers y flowers to the deity. Now she is left with no flowers in hand.

Q1. If Roopa leaves home with 30 flowers, the number of flowers she offers to each deity is
(a) 30 (b) 31 (c) 32 (d) 33

Q2. The minimum number of flowers that could be offered to each deity is
(a) 0 (b) 15 (c) 16 (d) Cannot be determined

Q3. The minimum number of flowers with which Roopa leaves home is
(a) 16 (b) 15 (c) 0 (d) Cannot be determined

Rough Work

■ Answer Grid

Q1	Q2	Q3
c	c	b

STORYLINE

As mentioned in the passage given, there are some conditions to be followed. Though one can solve these questions by back-solving (putting the options), it might take some time to get down to the solution. So, it is advisable that students should generate the equations by assuming the variables. Though this also takes its own time, but it makes the solution certain at the end.

Solutions to Q1–Q3

Roopa leaves home with x flowers. Then she dips these into the river and takes out $2x$ flowers.

Place of Worship	Number of Flowers Before Offering	Number of Flowers Offered	Number of Flowers Left
1	$2x$	Y	$2x - y$
2	$2\times(2x - y) = 4x - 2y$	Y	$(4x - 2y) - y = 4x - 3y$
3	$2\times(4x - 3y) = 8x - 6y$	Y	$(8x - 6y) - y = 8x - 7y$
4	$2\times(8x - 7y) = 16x - 14y$	Y	$(16x - 14y) - y = 16x - 15y$

Given that $16x - 15y = 0$, hence $16x = 15y$.

A1. Given that $x = 30 \Rightarrow 16 \times 30 = 15y$. Hence $y = 32$. Hence option (c) is the answer.

A2. Using $16x = 15y \Rightarrow x = \frac{15}{16}y$

The minimum value of y so that $\frac{15}{16}y$ is a whole number is 16. Therefore, 16 is the minimum number of flowers that can be offered. Hence option (c) is the answer.

A3. For y to be 16, value of $x = \frac{15}{16}y = 15$.

Hence, the minimum number of flowers with which Roopa leaves home is 15. Hence option (b) is the answer.

Goyal Family's Expenses Set 20

Direction for questions 1–4

Go through the data set given below and solve the questions based on it.

Pie charts given below present the expenses of Goyal family for two consecutive years 2013 and 2014. In 2012, data given is in percentage and in 2013, data given is in degrees.

Total expenses in 2013 = Rs 64,000 and total expenses in 2014 = Rs 96,000.

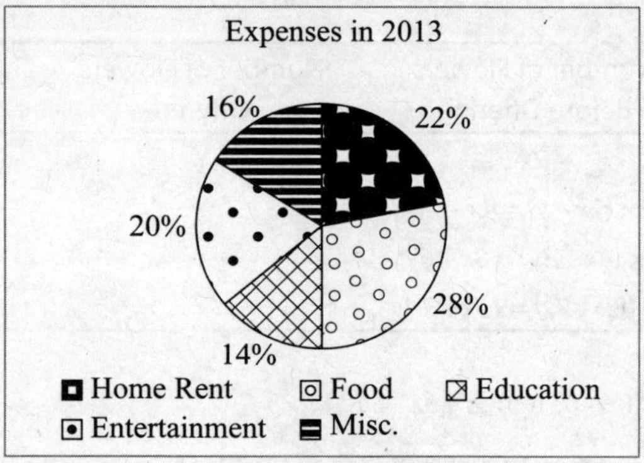

22% = Home rent and so on in clockwise direction.

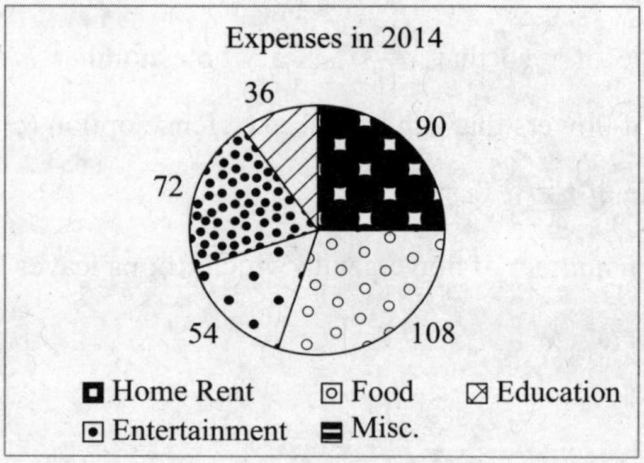

90 degrees = Home rent and so on in clockwise direction.

Q1. For how many sectors, spending in 2014 is less than the spending in 2013?

(a) 1 (b) 2 (c) 3 (d) 0

Q2. Which sector has seen the largest percentage increase in its spending in 2014 over 2013?

(a) Home rent (b) Food (c) Education (d) Entertainment

Q3. What is the percentage growth in the spending on Education in 2014 over 2013?

(a) 54% (b) 60% (c) 64% (d) 68%

Q4. Which sector has seen the largest absolute increase in spending in 2014 over 2013?
(a) Home rent (b) Food (c) Education (d) Entertainment

Rough Work

Answer Grid

Q1	Q2	Q3	Q4
a	a	b	b

STORYLINE

There are two set of data—One for 2013 and other for 2014. Pie chart given for 2013 has the data in percentage and pie chart given for 2014 has the data in degrees. To compare the data from one pie chart to other, we need to convert percentage into degrees or degrees into percentage. In my opinion, its always convenient to convert degrees into percentage data wherever required.

Solutions to Q1–Q4

Let us first convert data given in the 2nd pie chart into percentage:
360 degrees = 100% ⇒ 3.6 degrees = 1% ⇒ 18 degrees = 5%

	Expenses in 2014	In Percentage
Home rent	90	25%
Food	108	30%
Education	54	15%
Entertainment	72	20%
Misc.	36	10%

I would suggest you to write down the percentage value in the 2nd pie chart like given below:

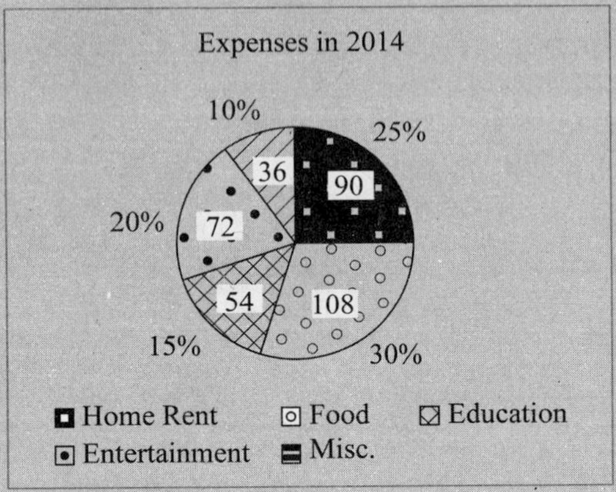

So finally, the tables look like this:

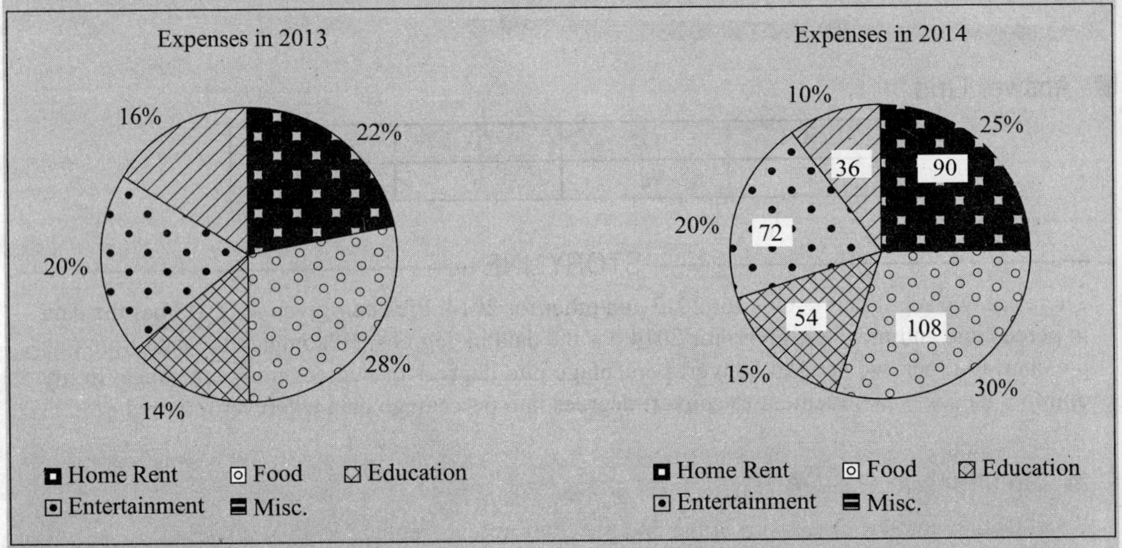

A1. Except in case of Misc., everywhere else percentage value has gone up. Since total value has gone up to Rs 96000 in 2014 from Rs 64000 in 2013, with an increase in percentage share, value of sectors like Home Rent, Food, Education, Entertainment are bound to rise.
Now we will verify for Misc. sector:
Value in 2013 = 16% of 64000 = 16% of 60000 + 16% of 4000 = 9600
Value in 2014 = 10% of 96000 = 9600
Hence value in 2013 > value in 2014.
Hence option (a) is the answer.

A2. To calculate the sectoral percentage change in 2014 over 2013, we can see that there are two factors responsible for it: (i) percentage increase in the share, (ii) total value has increased from Rs 64000 to Rs 96000.

Now, this second factor is constant for each of the sectors, hence it has equal impact over all the sectors. So we can neglect this and calculate only factor (i) to answer this question.

Percentage increase $= \frac{15}{16}y$

Entertainment has remained constant at 20% and Misc. has decreased. So we are not required to check for these two. It can be seen that 'Home rent' has seen the maximum change in the percentage value = 25% – 22% = 3%. To calculate percentage change, we are supposed to divide this by Initial value (=22). However, it can be further visualized that 3/22 is greater than 2/28 (Food sector) and 1/14 (education). Hence Home rent is the answer. Hence option (a) is the answer.

A3.

Percentage Share in 2013	Percentage Share in 2014
14%	15%

There are two methods of solving this question:

Method 1: Doing the actual calculation

Spending in 2013 = 14% of Rs 64000 = Rs 8960

Spending in 2014 = 15% of Rs 96000 = Rs 14400

$$\text{Percentage increase} = \frac{\text{increase}}{\text{initial value}} \times 100 = \frac{14400 - 8960}{8960} \times 100 = 60.7\%$$

Method 2: Successive percentage change

There are two factors responsible for the percentage increase—(a) rise in percentage share and (b) total spending has increased. We will calculate these one by one and use successive percentage increase to find out the net percentage increase.

(a) Rise in percentage share $= \frac{15 - 14}{14} \times 100 = 7\%$ (approx.)

(b) Rise in total spending $= \frac{96000 - 64000}{64000} \times 100 = 50\%$

Net percentage increase $= 7 + 50 + \frac{7 \times 50}{100} = 60.5\%$

Hence option (b) is the answer.

A4. Answer will be one of Home rent or Food.

Note: Just because increase in the percentage share for Home rent is 3%, and that of Food is 2%, we cannot say that absolute increase will be more in case of Home rent. It MAY be or MAY NOT be.

To be accurate, it is advisable that you go through the actual calculations.

	%	Total spend	Sector Spend	%	Total Spend	Sector Spend	Difference
Home Rent	0.25	96000	24000	0.22	64000	14080	9920
Food	0.3	96000	28800	0.28	64000	17920	10880

It can be seen that maximum absolute increase occurs in case of Food sector. Hence option (b) is the answer.

Two Years' Percentage Calculation

Set 21

Direction for questions 1–4

Go through the data set given below and solve the questions based on it.

Following table gives the percentage change in the sales revenue of five companies. Some of these data are missing and have been replaced by variable, *x* and *y*.

Companies	From 2010 to 2011	From 2011 to 2012
A	20%	–15%
B	8%	12%
C	14%	–6%
D	20%	x%
E	Y%	–8%

Q1. From 2010 to 2012, Company D's sales revenue has remained constant. What is the value of x?
(a) –20% (b) – 25% (c) –16.66% (d) Cannot be determined

Q2. If sales revenue of company A was Rs 100 in 2010, then what is its revenue in 2012?
(a) Rs 110 (b) Rs 102 (c) Rs 92 (d) Cannot be determined

Q3. Sales revenue of companies D and E are same in the year 2012. Which of the following between X and Y is greater?
(a) X (b) Y (c) X = Y (d) Cannot be determined

Q4. Which of the companies from A or B or C has the maximum sales volume?
(a) A (b) B (c) C (d) Cannot be determined

Rough Work

Answer Grid

Q1	Q2	Q3	Q4
c	b	d	d

STORYLINE

Question mentions the percentage change in the sales revenue of five companies. Understand that unless price per article is mentioned, we cannot find out the sales volume. To solve the questions ahead, you may be required to use (a) Successive Percentage change expression (b) Product Stability Ratio.

Solutions to Q1–Q4

A1.

Company	From 2010 to 2011	From 2011 to 2012
D	20%	%x

Method 1:

It is given that company D's revenue in 2010 is same as its revenue in 2012. So the increase (in amount) in 2011 shall be compensated (negatively) in 2012. Assume that the revenue in 2010 is Rs 100.

$$\text{Rs } 100 \xrightarrow{20\%\uparrow} \text{Rs } 120 (\text{in } 2011) \xrightarrow{x\%\downarrow} \text{Rs } 100 (\text{in } 2012)$$

$$\text{So percentage decrease} = \frac{\text{Decrease}}{\text{Initial Value}} \times 100 = \frac{20}{120} \times 100 = 16.66\%$$

Method 2: Use product stability ratio to find out the percentage decrease. Corresponding to percentage increase of 20% (1/5), percentage decrease should be (1/6) = 16.66%

[For a detailed account on Percentage, go through the book "Quantitative Aptitude and Data Interpretation for CAT" written by Nishit K Sinha].

Hence option (c) is the answer.

A2.

$$\text{Rs } 100 \xrightarrow{20\%\uparrow} \text{Rs } 120 (\text{in } 2011) \xrightarrow{15\%\downarrow} \text{Rs } 102 (\text{in } 2012)$$

A3. It is like population of two countries are same this year, and we have been asked to find out a relation between their populations two years back. It is not possible to determine. Hence option (d) cannot be determined is the answer.

A4. Information given in the question set is regarding the sales value. Hence any question regarding sales volume cannot be answered. Hence option (d) cannot be determined is the answer.

Respondents' Age

Set 22

Direction for questions 1–3

Go through the data set given below and solve the questions based on it.

The table below provides certain demographic details of 30 respondents who were part of a survey investigating the irregularities in NREGS. The demographic characteristics are: gender, number of children, and age of respondents. The first number in each cell is the number of respondents in that group. The minimum and maximum age of respondents in each group is given in brackets. For example, there are five female respondents with no children and among these five; the youngest is 34 years old, while the oldest is 49.

No. of children	Male	Female	Total
0	1(38,38)	5(34,49)	6
1	1(32,32)	8(35,57)	9
2	8(21,65)	3(37,63)	11
3	2(32,33)	2(27,40)	4
Total	12	18	30

Q1. The percentage of respondents aged less than 40 years is at least

(a) 10% (b) 16.67% (c) 20.0% (d) 30%

Q2. Given the information above, the percentage of respondents older than 35 can be at most

(a) 30% (b) 73.33% (c) 76.67% (d) 90%

Q3. The percentage of respondents that fall into the 35 to 40 years age group (both inclusive) is at least

(a) 6.67% (b) 10% (c) 13.33% (d) 26.67%

Rough Work

Answer Grid

Q1	Q2	Q3
d	c	c

STORYLINE

As explained in the opening statements, number of respondents can be counted up as per the requirement in the individual question.

Solutions to Q1–Q3

A1. We have to find out the minimum number of respondents less than age 40 years. Following is the number of respondents aged less than 40 years.

No. of children	Male	Female	Total
0	1	1	2
1	1	1	2
2	1	1	2
3	2	1*	3
Total	5	4	9

* Out of 2 females, one has to be of 40 years. Since question is asking for minimum number of respondent less than 40 years, we can count only one.

Percentage of minimum respondents aged less than 40 years $= \frac{9}{10} \times 100 = 30\%$

Hence option (d) is the answer.

A2. We have to find out the maximum number of respondents aged more than 35 years. Following is the number of respondents aged:

No. of children	Male	Female	Total
0	1	4	5
1	0	7	7
2	7	3	10
3	0	1	1
Total	8	15	23

Percentage of maximum respondents aged less more than 35years $= \frac{23}{30} \times 100 = 76.67\%$

Hence option (c) is the answer.

A3. Number of respondents in the given age bracket = 4

Hence percentage respondents $= \frac{4}{30} \times 100 = 13.33\%$

Hence option (c) is the answer.

Assembling the Table

Set 23

Direction for questions 1–3

Go through the information given below and solve the questions based on it.

ABC is a firm that deals with furniture. Manufacturing of table requires three levels of assembly. The finished table is at first level. The leg assembly and table-top are second level. The pieces that go into the leg assembly are at the third level which consists of short rails, long rails and legs. One unit of table requires one unit of table-top and one unit of leg assembly. One unit of leg assembly requires 2 units of short rails, 2 units of long rails and 4 units of legs. Orders are placed just in time to minimize storage.

The lead time for activities are (Lead time is waiting time required to complete one activity):

Parts	Weeks
Assemble table	1
Finished leg assembly	1
Purchase legs	1
Purchase short rails	1
Purchase long rails	1
Purchase table-top	2

The availability of parts at present:

Parts	Units
Finished table	50
Leg assembly	100
Legs	150
Short rails	50
Long rails	0
Table-top	50

Demand of finished tables:

Details	Week 4	Week 5	Week 6
Demand (units)	200	150	100

For delivering any table in subsequent weeks, demand of previous weeks must be met first. First demand occurs in week 4.

Q1. For meeting the demand of 200 units of finished table of week 4, when would the first order of table-tops be placed?

(a) Week 1 (b) Week 3 (c) Week 5 (d) None of these

Q2. Keeping in mind the inventory, how many legs should be ordered for meeting the demand of week 4 finished tables?

(a) 200 (b) 50 (c) 400 (d) 800

Q3. Keeping in mind the inventory, when and how many units of short rail would be placed for meeting the demand of finished table of week 6?

(a) 100 units in week 1 (b) 200 units in week 3
(c) 300 units in week 6 (d) None of these

Rough Work

Answer Grid

Q1	Q2	Q3
a	b	d

STORYLINE

There are three levels of work in table making:

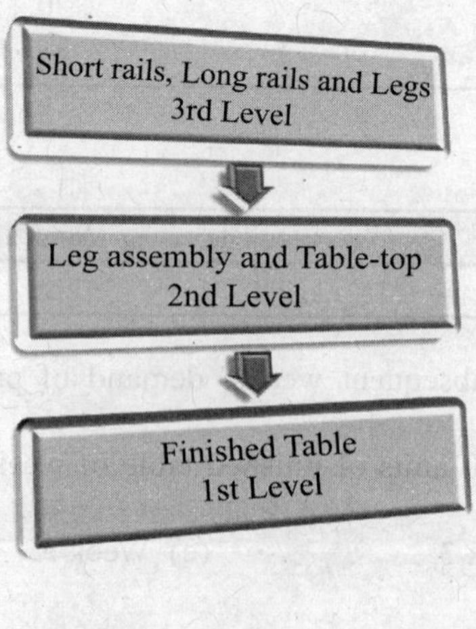

For the sake of understanding the terms used (Short rail and long rail), have a look at the following picture (though we really don't need to understand this to solve the questions):

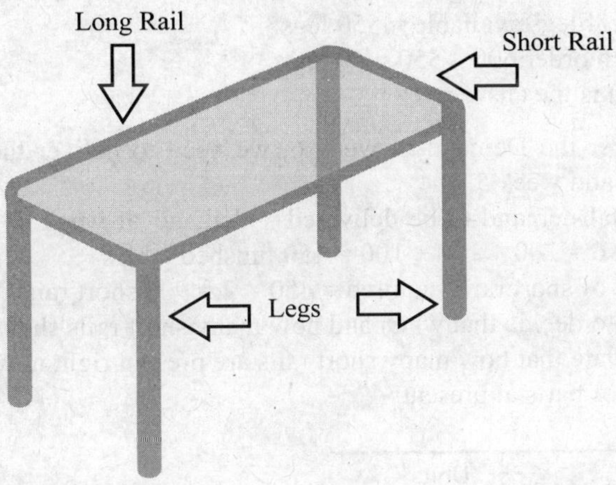

Solutions to Q1–Q3

A1. We need to have 200 units of finished tables in week 4.
For 200 tables, we need to have the following:
200 units of tabletop and 200 units of leg assembly.
200 units of leg assembly = 200 × (2 units of short rails, 2 units of long rails and 4 units of legs) = 400 units of long rails + 400 units of short rails + 1600 legs.
We already have 50 finished tables and 50 table top with us. So we need to order 100 tabletops.
Lead time involved related to table tops are – Purchase table tops (2 weeks) and assemble table (1 week). So total lead time = 3 weeks.
Since tables are to delivered in week 4, hence we need to place the order 3 weeks back (keeping a lead time of 3 week). Hence table top shall be ordered in 4 – 3 = 1st week.
Hence option (a) is the answer.

A2. Total demand of finished tables in week 4 = 200 tables.
Net availability right now:

Parts	Units
Finished Table	50
Leg assembly	100
Legs	150

We already have 50 finished table, so we need to manufacture 150 tables.
For this, we need to have 150 × 4 = 600 legs.

Out of this, following number of legs are available:
100 leg assembly contains = 100 × 4 = 400 legs.
Spare legs available = 150 legs.
So total number of legs available = 550 legs.
Hence we need to order 600 – 550 = 50 legs.
Hence option (b) is the answer.

A3. Before we deliver the Demand of week 6, we need to deliver the demands of previous weeks – week 4 and week 5.
So, in effect, total demand to be delivered = demand of week 4 + demand of week 5 + demand of week 6 = 200 + 150 + 100 = 450 finished tables
So, total number of short rails required = 450 × 2 = 900 short rails
We are required to decide that when and how many short rails shall be ordered?
Let us first calculate that how many short rails are present right now.
The availability of parts at present:

Parts	Units
Finished table	50
Leg assembly	100
Legs	150
Short rails	50
Long rails	0
Table-top	50

Total number of short rails present = 50 × 2 (in Finished Table) + 100 × 2 (in Leg Assembly) + 50 (Short rails) = 250 Short rails.
Hence total number of short rails required to be ordered = 900 – 250 = 650 short rails.
Since none of the options contain 650 short rails in it, we are not required to check further that when is it required to be ordered.
Hence option (d) None of these is the answer.

Beverage Index

Set 24

Direction for questions 1–4

Go through the data set given below and solve the questions based on it.

The line chart depicts the price trend of beverages (Rs/kg) from 2000 to 2009.

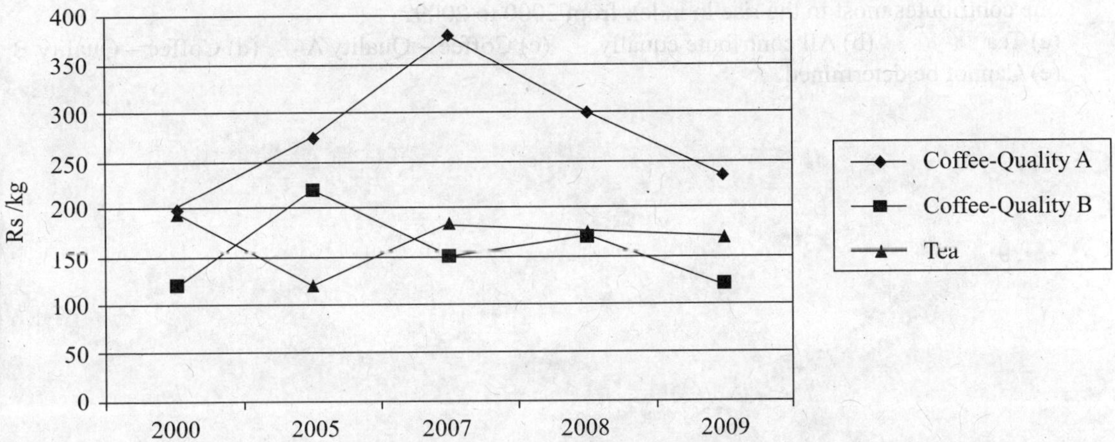

Following table provides the price indices for agricultural commodities in the years 2000 to 2009. For the sake of calculation, base year is considered as the year 2000. In the year 2000, unit price of the items mentioned = 100.

Price indices for agricultural Commodities					
Item	2000	2005	2007	2008	2009
1. Beverages	100	127	158	135	104
2. Food	100	98	107	101	85
3. Raw Materials	100	113	105	84	85

Assume that the index for beverage is based on the 3 beverages for which the price trend is shown in the line chart given at the top.

Q1. In 2009, what is the quantity of tea that can be bought for Rs 2,000?

(a) 10.8 kg (b) 5.88 kg (c) 11.7 kg (d) 12.8 kg (e) 11 kg

Q2. During which period was the change in the price (Rs/kg/years) the lowest for coffee quality A?

(a) 00–0 (b) 05–07 (c) 07–08 (d) 08–09 (e) Cannot be determined

Q3. For which of the three beverages is the change in price between 00 and 09 not in conformity with the trend for beverage price index?
I. Coffee – Quality A
II. Coffee – Quality B
III. Tea
(a) III only (b) I and II (c) I and II (d) II and III
(e) II only

Q4. Assuming that equal weightage is given to each of three beverages in calculating the index, which one contributes most to the rise in index from 2000 to 2009?
(a) Tea (b) All contribute equally (c) Coffee – Quality A (d) Coffee – Quality B
(e) Cannot be determined

Rough Work

Answer Grid

Q1	Q2	Q3	Q4
b	a	d	c

STORYLINE

Line charts are typically used for trend analysis for chronological data. A clear trend of price change can be seen through the line chart—with a word of caution because gap in number of years between two data point is not same. Table given below presents the index value for Beverages, based upon the data from line chart, and two other items. As stated in the first line, you may expect questions which ask to compare the trends of different items across the years.

Solutions to Q1–Q4

A1. This question needs a simple calculation. Price per kg of tea in 2009 = Rs 170/kg.

Hence quantity that can be bought for Rs 1000 $= \frac{1000}{170} =$ Rs 5.88/kg

Hence option (b) is the answer.

A2. Question is asking for change, and not increase or decrease. So we are required to work upon with magnitude only, and sign (positive or negative) is not important.

Year Span	Total Change	Total Years	Change/year
2000–2005	75	5	15
2005–2007	105	2	52.5
2007–2008	80	1	80
2008–2009	65	1	65

It is obvious that change/year is minimum in the year 2000–2005. Hence option (a) is the answer.

A3. Index has the following pattern: Increase – Increase – Decrease – Decrease

Only Coffee Quality A shows this trend.

Hence option (d) is the answer.

A4. Beverages Index in the year 2000 = 100

Beverages Index in the year 2009 = 104. So there is a rise in index.

It can be observed that Tea price has remained same and price of Coffee Quality A has gone down in 2009 over 2000.

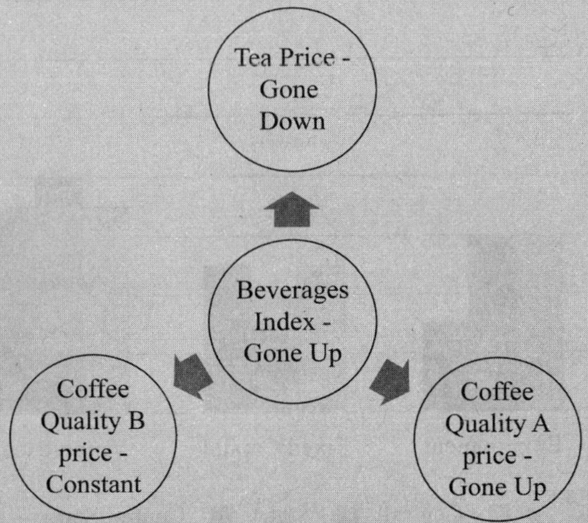

Only the price of Coffee Quality A has gone up. So it can be concluded that Coffee quality A has the maximum impact on the beverages index and hence contributes most.

Hence option (c) is the answer.

Sectoral Data of Factories Set 25

Direction for questions 1–4

Go through the chart given below and solve the questions based on it.

Following grid presents the percentage composition of five verticals under different sectors viz., Government, Public, Private, Joint.

Sector	Factories	Employment	Fixed Capital	Variable Cost	Value Added
Government	18	15	14	22	25
Public	12	8	6	19	8
Private	55	65	72	54	62
Joint	15	12	8	5	5
Total	100	100	100	100	100

For example, of all the factories, 18% are owned by Government.

Further, Government sector can be further broken down into three sub-sectors—Central, State, Central/State. Following bar chart gives a break-up of the Government Sector data (as given in Table above):

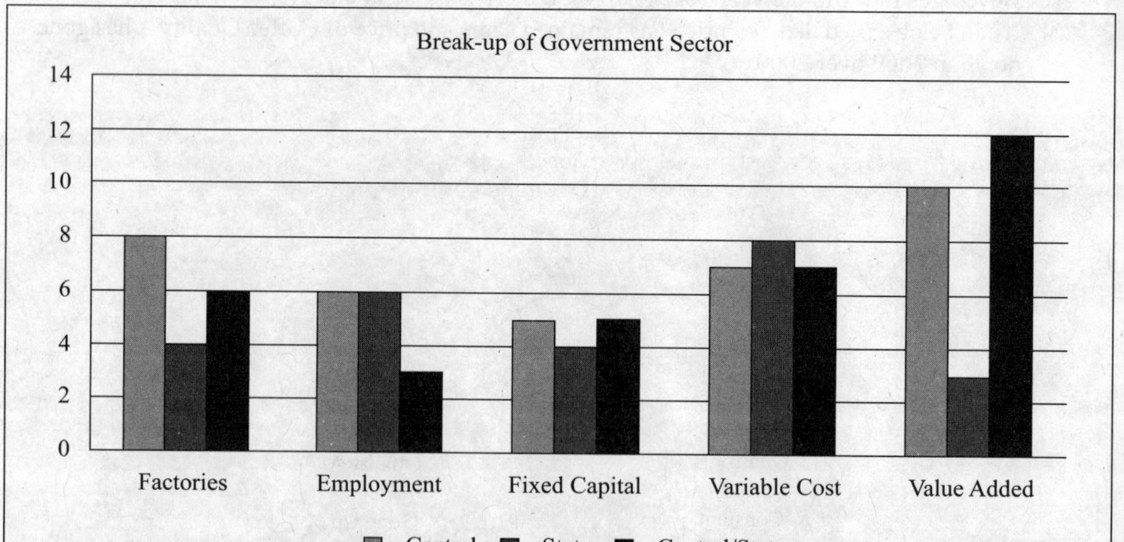

Q1. If the total work force was 76 million whereas the total value added was Rs 225 million, then which of the following had the maximum value addition per worker?
(a) Central (b) State (c) Central/State (d) Public

Q2. Which of the following sectors has the maximum fixed capital invested per factory?
(a) Central (b) State (c) Central/State (d) Public

Q3. If the variable is proportional to the number of employees and the production per employee, then for which of the following is the production highest?
(a) Government (b) Private (c) Joint (d) Public

Q4. If the government has a fixed capital of $200 million in the Iron & Steel industry, which corresponds to 20.012% of its total investment as fixed capital, then how much did the government invest (in Rs million) in Maruti Udyog Ltd which forms 25% of the investment in the joint sector?(1US $ = Rs 45)
(a) 7860 (b) 2500 (c) 143 (d) 145

Rough Work

Answer Grid

Q1	Q2	Q3	Q4
c	b	b	a

STORYLINE

Table provides percentage share (or composition) of factories, employment, fixed capital, variable cost and value added among four sectors viz., government sector, public sector, private sector and joint sector.

Key point here is to understand that Government sector is broken down into three sub-sectors. For all the data given in Government sector, corresponding data for sub-sectors is provided in the bar chart.

Solutions to Q1–Q4

A1. Value addition per worker = $\frac{\text{Value Added}}{\text{No. of Employes}}$

Total Value added = Rs 225 million and Total work force = 76 million.

We go by the options and calculate value added per worker for all the given options:

For Central = $\frac{10\% \text{ of Rs } 225 \text{ million}}{6\% \text{ of } 76 \text{ million}}$

For State = $\frac{8\% \text{ of Rs } 225 \text{ million}}{6\% \text{ of } 76 \text{ million}}$

For Central/State = $\frac{12\% \text{ of Rs } 225 \text{ million}}{8\% \text{ of } 76 \text{ million}}$

For Public = $\frac{8\% \text{ of Rs } 225 \text{ million}}{8\% \text{ of } 76 \text{ million}}$

Calculation Short-cut:

Inserted of calculating the real value in the above four options, what we can do is quick ratio comparison. We can observe that $\frac{\text{Rs } 225 \text{ million}}{76 \text{ million}}$ is present in all the ratios. Let us assume $\frac{\text{Rs } 225 \text{ million}}{76 \text{ million}} = \text{N}$.

So the four ratios become like the following:

For Central = $\frac{10\% \text{ of Rs } 225 \text{ million}}{6\% \text{ of } 76 \text{ million}} = \frac{10\%}{6\%}\text{N}$

For State = $\frac{8\% \text{ of Rs } 225 \text{ million}}{6\% \text{ of } 76 \text{ million}} = \frac{8\%}{6\%}\text{N}$

For Central/State = $\frac{12\% \text{ of Rs } 225 \text{ million}}{8\% \text{ of } 76 \text{ million}} = \frac{12\%}{8\%}\text{N}$

For Public = $\frac{8\% \text{ of Rs } 225 \text{ million}}{8\% \text{ of } 76 \text{ million}} = \frac{8\%}{8\%}\text{N}$

It can be now clearly seen that the ratio is largest in case of Central/State. Hence option (c) is the answer.

In such questions, we can do the proceeding in a nutshell like the following:

Value added per worker for various sectors will be proportional to following ratios:

Central	10/6
State	3/6
Central/state	12/3
Public	8/8

A2. Using the method given in A1:
Fixed capital per factory will be proportional to the following ratios:

Central	5/8
State	4/4
Central/state	5/6
Public	6/12

Hence option (b) is the answer.

A3. If the variables are proportional to the number of employees and the production per employee, then we can say that the variable cost is directly proportional to the production. Since the variable cost is maximum for private sector, it will also have the maximum production.
Hence option (b) is the answer.

A4. Since options are not so close, we can take 20.012% = 20%
If 20% of government fixed capital is equal to \$200 million => Total government sector fixed capital = \$ 1000 m which is equivalent to 14% of the total fixed capital. Now, 25% of government investment in joint sector i.e., 25% of the 8% of the total=2% of total fixed capital will be given by: $\frac{100 \times 2}{14} = \$\ 143$ million
Value in Rs is given by: 143 × 55 = Rs 7865 million
Hence option (a) is the answer.

Idli Vada

Set 26

Direction for questions 1–3

Go through the information given below and solve the questions based on it.

Five friends meet every morning at Sree sagar restaurant for an idli-vada breakfast. Each consumes a different number of idli and vada. The number of idli's consumed are 1, 4, 5, 6 and 8, while the number of vadas consumed are 0, 1, 2, 4 and 6. Below are some more facts about who eats what and how much:

The number of vadas eaten by Ignesh is three times the number of vadas consumed by the person who eats four idlis.

Three persons, including the one who eats four vadas eat without chutney.

Sandeep does not take any chutney.

The one who eats one idli does not eat any vadas or chutney. Further, he is not Mukesh.

Daljits eats idli with chutney and also eats vada.

Mukesh, who does not take chutney, eats half as many vadas as the person who eats twice as many idlis as he does.

Bimal eats two more idlis than Ignesh, but Ignesh eats two more vadas than Bimal.

Q1. Which of the following statement is true?

(a) Daljit eats 5 idlis
(b) Ignesh eats 8 idlis
(c) Bimal eats 1 idli
(d) Bimal eats 6 idlis.

Q2. Which of the following statement is true?

(a) Sandeep eats 2 vadas.
(b) Mukesh eats 4 vadas.
(c) Ignesh eats 6 vadas.
(d) Bimal eats 2 vadas.

Q3. Which of the following statements is true?

(a) Mukesh eats 8 idlis and 4 vada but not chutney
(b) The person who eats 5 idlis and 1 vada does not take chutney.
(c) The person who eats equal number of vadas and idlis also takes chutney.
(d) The person who eats 4 idlis and 2 vada also takes chutney.

Rough Work

Answer Grid

Q1	Q2	Q3
a	c	c

STORYLINE

This is a classic analytical reasoning question involving data—you are expected to match the variables (Name of person – Number of idlis and vadas eaten) as per the conditions given.

I have put this set in this book because one is required to crack the data (number of idlis/vadas eaten) using the concept of ratio to solve this set.

Solutions to Q1–Q3

To solve such questions, it is advisable to draw table using the information given—one by one:

Using point (i):

(a) Number of vadas eaten by Ignesh is three times the number of vadas consumed by somebody. Hence number of vadas eaten by Ignesh should be 3 times of a number. Only possibility here = 6

(b) We also know that the person **who eats four idlis eats two vadas**. (So that Ignesh can eat three times the number of vadas eaten by this person).

	Idlis					Vadas				
Name	1	4	5	6	8	0	1	2	4	6
Ignesh						No	No	No	No	Yes
Sandeep										No
Mukesh										No
Daljit										No
Bimal										No

Using point (ii): This does not give any concrete information. So we move on to the next point (iii).

Using point (iii): This does not give any concrete information. So we move on to the next point (iv).

Using point (iv):

(a) **Person who eats one idli eats zero vada.**

(b) He is not Mukesh. So Mukesh cannot eat 0 (zero) vadas.

So we can fill the relevant blocks in the grid.

	Idlis					Vadas				
Name	1	4	5	6	8	0	1	2	4	6
Ignesh						No.	No	No	No	Yes
Sandeep										No
Mukesh	No					No				No
Daljit										No
Bimal										No

Using point (v):

(a) Daljit eats both idli and vada. So we will put a NO in 0 vadas (zero vadas).

	Idlis					Vadas				
Name	1	4	5	6	8	0	1	2	4	6
Ignesh						No	No	No	No	Yes
Sandeep										No
Mukesh	No					No				No
Daljit						No				No
Bimal										No

Using Point (vi):

(a) Number of vadas eaten by Mukesh = Half of the number of vadas eaten by somebody (let us call him X).

This gives us two possibilities – either Mukesh eats 2 vadas or 1 vada. In any case, Mukesh cannot eat 4 vadas.

(b) Number of idlis eaten by Mukesh = Half of the number of idlis eaten by somebody (the same guy X).

In this case, the only possibility is – Mukesh eats 4 idlis and that 'somebody X' eats 8 idlis.

We will fill these information in the relevant blocks of the grid:

	Idlis					Vadas				
Name	1	4	5	6	8	0	1	2	4	6
Ignesh		***No***				No	No	No	No	Yes
Sandeep		***No***								No
Mukesh	No	***Yes***	***No***	***No***	***No***	No			***No***	No
Daljit		***No***				No				No
Bimal		***No***								No

(New blocks have been filled using ***bold and italics*** font).

Using point (vii):

(a) Bimal eats two more idlis than Ignesh ⇒ Only possible option for this to happen is: Bimal = 8 idlis and Ignesh = 6 idlis.

	Idlis					Vadas				
Name	1	4	5	6	8	0	1	2	4	6
Ignesh	*No*	**No**	*No*	*Yes*	*No*	**No**	**No**	**No**	**No**	**Yes**
Sandeep		No		***No***	***No***					No
Mukesh	No	Yes	No	***No***	***No***	No			No	No
Daljit		No		***No***	***No***	No				No
Bimal	***No***	No	***No***	***No***	***Yes***					No

(New blocks have been filled using ***bold and italics*** font).

(b) Ignesh eats two more vadas than Bimal ⇒ Ignesh eats 6 vadas, so, Vimal eats 4 vadas.

	Idlis					Vadas				
Name	1	4	5	6	8	0	1	2	4	6
Ignesh	No	No	No	Yes	No	No	No	No	No	Yes
Sandeep		No		No	No				***No***	No
Mukesh	No	Yes	No	No	No	No			***No***	No
Daljit		No		No	No	No			***No***	No
Bimal	No	No	No	No	Yes	***No***	***No***	***No***	***Yes***	No

(New blocks have been filled using ***bold and italics*** font).

Now, we have used all the direct statements. Its time to join two or more statements together to fill in the remaining blocks.

Using (i)b - the person **who eats four idlis eats two vadas** ⇒ Since Mukesh **eats four idlis,** Mukesh eats two vadas.

	Idlis					Vadas				
Name	1	4	5	6	8	0	1	2	4	6
Ignesh	No	No	No	Yes	No	No	No	No	No	Yes
Sandeep		No		No	No			***No***	No	No
Mukesh	No	Yes	No	No	No	No	***No***	***Yes***	No	No
Daljit		No		No	No	No		***No***	No	No
Bimal	No	No	No	No	Yes	No	No	No	Yes	No

(New blocks have been filled using ***bold and italics*** font).

Now, for Daljit in Vadas, only one block is empty. This is shown in grey colour.

	Idlis					Vadas				
Name	1	4	5	6	8	0	1	2	4	6
Ignesh	No	No	No	Yes	No	No	No	No	No	Yes
Sandeep		No		No	No			***No***	No	No
Mukesh	No	Yes	No	No	No	No	***No***	***Yes***	No	No
Daljit		No		No	No	No	??	***No***	No	No
Bimal	No	No	No	No	Yes	No	No	No	Yes	No

This further fills in the grey cells. Other cells filled in this way are shown in grey colour:

	Idlis					Vadas				
Name	1	4	5	6	8	0	1	2	4	6
Ignesh	No	No	No	Yes	No	No	No	No	No	Yes
Sandeep		No		No	No		No	***No***	No	No
Mukesh	No	Yes	No	No	No	No	***No***	***Yes***	No	No
Daljit		No		No	No	No	Yes	***No***	No	No
Bimal	No	No	No	No	Yes	No	No	No	Yes	No

This leaves us with only one option for Sandeep: Number of vadas eaten = 0.

	Idlis					Vadas				
Name	1	4	5	6	8	0	1	2	4	6
Ignesh	No	No	No	Yes	No	No	No	No	No	Yes
Sandeep		No		No	No	**Yes**	No	No	No	No
Mukesh	No	Yes	No	No	No	No	No	Yes	No	No
Daljit		No		No	No	No	Yes	No	No	No
Bimal	No	No	No	No	Yes	No	No	No	Yes	No

(Shown in grey cell).
Using (iv)a - **Person who eats one idli** eats **zero Vada.** Since **Sandeep** eats **zero vada,** he eats **one idli.**

	Idlis					Vadas				
Name	1	4	5	6	8	0	1	2	4	6
Ignesh	No	No	No	Yes	No	No	No	No	No	Yes
Sandeep	***Yes***	No	***No***	No	No	Yes	No	No	No	No
Mukesh	No	Yes	No	No	No	No	No	Yes	No	No
Daljit	***No***	No	***Yes***	No	No	No	Yes	No	No	No
Bimal	No	No	No	No	Yes	No	No	No	Yes	No

(new blocks have been shown in grey color).

This leaves only one option for Sandeep – 5 idlis (Shown in above table).

Finally, we arrive at the following table:

	Idlis					Vadas				
Name	1	4	5	6	8	0	1	2	4	6
Ignesh	No	No	No	Yes	No	No	No	No	No	Yes
Sandeep	Yes	No	No	No	No	Yes	No	No	No	No
Mukesh	No	Yes	No	No	No	No	No	Yes	No	No
Daljit	No	No	Yes	No	No	No	Yes	No	No	No
Bimal	No	No	No	No	Yes	No	No	No	Yes	No

Summary Chart:

	Idli	Vada
Ignesh	6	6
Sandeep	1	0
Mukesh	4	2
Daljit	5	1
Bimal	8	4

A1. Daljit eats 5 idlis. Hence option (a) is the answer.

A2. Ignesh eats 6 vadas. Hence option (c) is the answer.

A3. Use statement (ii) – three persons eat without chutney. Hence option (c) is the answer.

Chandigarh to Delhi

Set 27

Direction for questions 1–5

Go through the data set given below and solve the questions based on it.

There are two bus services available to travel from Chandigarh (CHD) to Delhi—one of Pat Roadways and the other of Rajan Roadways. The distance between the two places is 250 km. Each o these buses stop at four locations on the way from Chandigarh to Delhi. These locations are Ambal (AB), Kurukshetra (KU), Saharanpur (SH) and Panipat (PAN). The arrival and departure times of th two buses are given below in the table:

	CHD (dep)	AB (arr)	AB (dep)	KU (arr)	KU (dep)	SH (arr)	SH (dep)	PAN (arr)	PAN (dep)	Delhi (arr)
Patel Roadways	09:30	10:40	10:42	11:30	11:35	12:10	12:12	13:40	13:42	14:30
Rajan Roadways	14:30	15:40	15:42	16:20	16:23	17:00	17:15	18:15	18:18	19:00

Note *Arr stands for arrival and Dep stands for departure.*

Q1. What is the difference between the speeds of faster and slower bus (including stoppages)?
(a) 3.33 km/h (b) 5.55 km/h (c) 6.66 km/h (d) 7.77 km/h

Q2. Between which stretches is the ratio of the speeds of the two bus services 1:1?
(a) AB to KU (b) CHD to AB (c) SH to PAN (d) Cannot be determined

Q3. What is the percentage difference between the maximum speed of Rajan Roadways bus and th minimum speed of Patel Roadways bus?
(a) 12% (b) 19% (c) 26% (d) Cannot be determined

Q4. In which stretches of the journey is the speed of Patel Roadways bus minimum (Assume distanc between the stops to be equal)?
(a) CHD to AB (b) AB to KU (c) KU to SH (d) SH to PAN

Q5. If the average speed is calculated keeping in mind only the running time of the bus, what woul be the percentage difference in time taken by the Patel Roadways bus, if its speed is decreased b 20%?
(a) 20% (b) 25% (c) 33.33% (d) 42.5%

Rough Work

Answer Grid

Q1	Q2	Q3	Q4	Q5
b	b	d	d	b

STORYLINE

Based upon the concepts of Time, Speed and Distance (TSD), this set requires you to use the fundamentals and basic formula Distance = Time × Speed. There are five stretches of the whole journey from Chandigarh to Delhi – Chandigarh to Ambala, Ambala to Kurukshetra, Kurukshetra to Saharanpur, Saharanpur to Panipat and Panipat to Delhi. Data to calculate the time taken in any stretch is given for two bus services—Patel Roadways and Rajan Roadways. Since total distance is given (=250 km), speed of the bus services for the whole journey can be calculated. Though you should keep in your mind that NOT all the stretches are of equal length, so despite time taken in any stretch is given, speed in any particular stretch cannot be calculated.

Solutions to Q1–Q5

A1. Total time taken by Patel Roadways bus = 5 hours

Average speed of Patel Roadways bus = $\frac{250}{5}$ = 50 km/h

Total time taken by Rajan Roadways bus = 4.5 hours

Average speed of Patel Roadways bus = $\frac{250}{4.5}$ = 55.55 km/h

[Instead of calculating $\frac{250}{4.5}$, you should calculate $\frac{500}{9} = 55.55$ kmph]

Hence slower by 5.55 km/h. Hence, option (b) is the answer.

A2. Though individual speed cannot be calculated, ratio of speeds can be calculated. It can be seen that time taken by both bus services is same between CHD to AB.
Hence option (b) is the answer.

A3. Distance between any two stretches is not given. So, despite time taken between any two stretches can be calculated by taking the difference, we cannot calculate the speed.
Hence answer is option (d) cannot be determined.

A4. This question allows us to have the assumption that all the stretches are of equal length. We can answer this question without calculating the actual speed. We just have to see that in which stretch time taken to travel is maximum. More is the time taken, less is the speed. Time taken to travel from SH to PAN is 1 hour 28 min, which is the highest. Hence, Speed is lowest in this stretch.
Hence, option (d) is the answer.

A5. We don't need to go through the table given to solve this question. This question is based upon the simple application of ratio/percentage in TSD.

If speed decreases by 20% then time increases by 25% ($= \frac{1}{0.8} = 1.25$).
Hence, option (b) is the answer.

Chunmun's Favourite Fruits

Set 28

Direction for questions 1–4

Answer the questions on the basis of the information given below.

Chunmun likes following six different types of fruits—Mangoes, Apples, Oranges, Peaches, Guavas and Litchis. Following table provides information about the minimum and the maximum price of each of the mentioned fruits. The table also provides information about the number of fruits of each type and the average price of each fruit.

Assume that the price of each fruit is an integer. There is only one mango that has the minimum price and there is only one mango that has the maximum price. The same holds true for each type of fruit.

	Minimum Price	Maximum Price	Average Price	Number of Fruits
Mango	20	28	23	16
Apple	40	50	46	24
Orange	4	8	7	40
Peach	16	26	20	34
Guava	3	8	5	28
Litchi	10	18	16	30

Q1. Find the approximate average price of fruits with Chunmun.
(a) Rs 18 (b) Rs 20 (c) Rs 23 (d) Rs 15

Q2. The number of mangoes whose price is more than Rs 21 is at least
(a) 3 (b) 4 (c) 5 (d) 6

Q3. Given that the price of four peaches with Chunmun is Rs 22 each and there are exactly 10 peaches whose price is Rs 18. The number of peaches with Chunmun whose price is at least Rs 25 is at most
(a) 13 (b) 12 (c) 11 (d) 9

Q4. Total number of guavas with Chunmun whose price is less than the average price of the guavas with Chunmun is at least
(a) 17 (b) 18 (c) 19 (d) 20

Rough Work

Answer Grid

Q1	Q2	Q3	Q4
a	d	d	b

Solutions to Q1–Q4

A1. Total price of fruits with Chunmun is 23 × 16 + 46 × 24 + 7 × 40 + 20 × 34 + 5 × 28 + 16 × 30 = Rs 368 + Rs 1104 + Rs 280 + Rs 680 + Rs 140 + Rs 480 = Rs 3052. Total number of fruits with Chunmun is 16 + 24 + 40 + 34 + 28 + 30 = 172. Average price of fruits with Chunmun = (3052/172) = Rs 17.74

Closest value given in the options is Rs 18. Hence option (a) is the answer.

A2. The total price of mangoes with Chunmun is 16 × 23 = Rs 368.

Total price of 14 mangoes = 368 – (20 + 28) = Rs 320

Out of the given 14 mangoes assume that maximum number of mangoes have price equal to Rs 21.

The maximum price of any mango out of these 14 mangoes is Rs 27

So, 9 × 21 + 1 × 23 + 4 × 27 = Rs 320

Therefore, the number of mangoes whose price is more than Rs 21 is 1 + 4 + 1 = 6.

Hence option (d) is the answer.

A3. The price of the 18 peaches with Chunmun whose price is not known to us = X

X = 20 × 34 – (16 + 26 + 4 × 22 + 10 × 18) = Rs 370

Assume that maximum number of peaches have price equal to Rs 17

370 = 25 × 8 + 17 × 10. Number of peaches with Chunmun whose price is at least Rs 25 is at most 8 + 1 = 9.

Hence option (d) is the answer.

A4. The total price of the guavas with Chunmun is 5 × 28 = 140
The total price of the 26 guavas whose price is not known is 140 – (3 + 8) = Rs 129.
Assume maximum number of guavas have price equal to Rs 7
129 = 8 × 7 + 1 × 5 + 17 × 4. Therefore, the number of guavas with Chunmun whose price is less than the average price of the guavas with Chunmun is at least 17 + 1 = 18.
Hence option (b) is the answer.

Population Data

Set 29

Direction for questions 1–4

Go through the data set given below and solve the questions based on it.

Following table gives the data regarding population of selected States and Union Territories (UTs) of India. Same data is given for India also.

State/ Union Territory	India/States/ Union Territory #	Sex ratio (Females per 1,000 Males)					
		Total population		Child population (less than 7 years)		Population (7 years and above)	
		2001	2011	2001	2011	2001	2011
C1	C2	C3	C4	C5	C6	C7	C8
INDIA	INDIA	933	940	927	914	934	944
01	Jammu & Kashmir	892	883	941	859	884	887
02	Himachal Pradesh	968	974	896	906	980	983
03	Punjab	876	893	798	846	888	899
04	Chandigarh	777	818	845	867	767	812
05	Uttarakhand	962	963	908	886	973	975
06	Haryana	861	877	819	830	869	885
07	NCT of Delhi	821	866	868	866	813	866
08	Rajasthan	921	926	909	883	923	935
09	Uttar Pradesh	898	908	916	899	894	910
10	Bihar	919	916	942	933	914	912
11	Sikkim	875	889	963	944	861	883
12	Arunachal Pradesh	893	920	964	960	878	913
13	Nagaland	900	931	964	944	890	929
14	Manipur	974	987	957	934	977	995
15	Mizoram	935	975	964	971	930	976

Q1. How many states/UTs have shown the same trend of increase or decrease as shown by India across C3 to C8?

(a) 6 (b) 7 (c) 8 (d) 9

Q2. What percentage of total population of Bihar in 2011 are females?
(a) 51.8% (b) 47.8% (c) 49.1% (d) 50.2%

Q3. What percentage of total population of India in 2001 is of children less than 7 years age?
(a) 16.66% (b) 14.28% (c) 11.11% (d) Cannot be determined

Q4. For how many States or Union Territories in 2011, percentage of children aged less than 7 years is atleast 50% of the total population of the that state or union territory?
(a) 1 (b) 2 (c) 6 (d) 0

Answer Grid

Q1	Q2	Q3	Q4
c	b	b	a

STORYLINE

Table presents the population data of Indian and 15 states/UTs for two years—2001 and 2011. Further, table also gives the sex ratio of these states under two headings—Age less than 7 years and 7 years and above. Since a unified data is also given, this gives rise to the case of weighted average.

Solutions to Q1–Q4

A1. States that show the same trend are—Uttarakhand, NCT of Delhi, Rajasthan, Uttar Pradesh, Sikkim, Arunachal Pradesh, Nagaland and Manipur = 8 states. Hence option (c) is the answer.

A2. For every 1000 males, there 916 females in Bihar. Using this we can find out the percentage of females in the total population of Bihar.
Actual Calculation:

$$\text{Percentage of females} = \frac{916}{1916} \times 100 = 47.8\%$$

Approximation Short-cut:
Since number of males is more than number of females, hence percentage of males will be more than the percentage of females. So percentage of males > 50% and percentage of females < 50%. Due to this, we can eliminate options (d) and (a).
Total population of Bihar = 916 x + 1000 x = 1916 x

We are required to calculate: Percentage of females = $\frac{916}{1916}$ [we have removed x from the numerator and denominator]

Knowing the fact that females' population is less than 50% and the remaining options hovering around 47%–49%, we benchmark the whole data around 50%.

$$50\% \text{ of } 1916 = \frac{1}{2} 1916 = 958$$

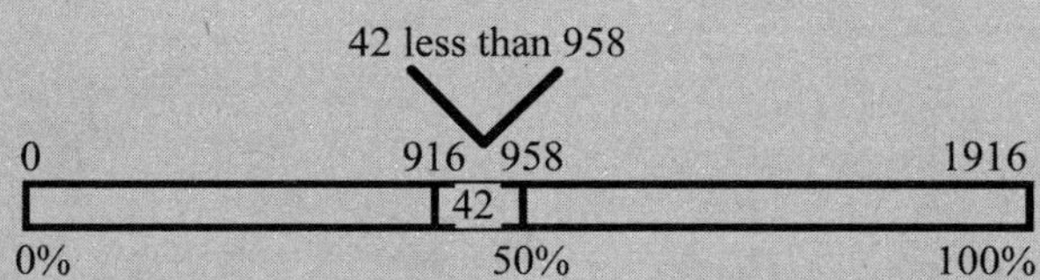

1% of 1916 = 19.16 = Approx. 20
So, 42 is approx. 2.1% of 1916. Hence 916 is 2.1% less than 50% = 47.9%. Closest option is option (b) 47.9%. Hence option (b) is the answer.

A3. This is a direct question on Weighted Average concept. We can solve this question using Criss-cross solving method of alligation:

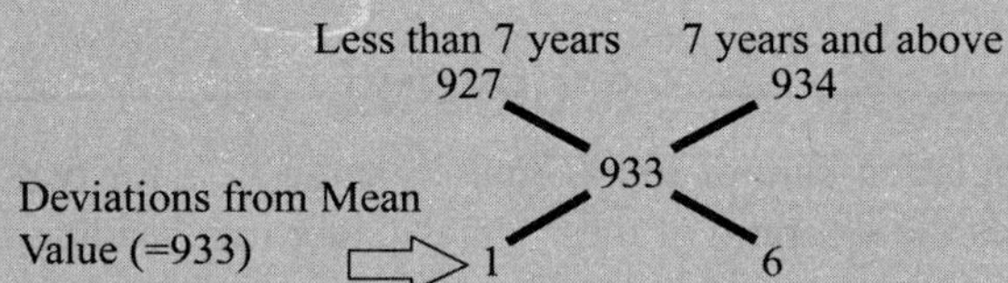

$$\text{So, } \frac{\text{Children of age less than 7years}}{\text{Aged 7 years and above}} = \frac{1}{6}$$

Hence percentage of children of age less than 7 years out of total population

$$= \frac{1}{7} \times 100 = 14.28\%. \text{ Hence option (b) is the answer.}$$

A4. Using the method of alligation (as shown in A3), we can simply conclude that:
For all the states/ UTs, in which deviation of "child population less than 7 years" from "Total Population" is less than the deviation of 'child population above 7 years' from Total population, percentage of "child population less than 7 years" will be 50% or more.
Now we just have to see that for how many states/UTs, "child population less than 7 years" is more than the value given for the whole state/UT.

State/ UT	India/States/ Union Territory #	Sex ratio (females per 1,000 males)			
		Total population	Child population (less than 7 years)	Population (7 years and above)	|Column (8) – Column (4)|> |Column (6) – Column (4)|?
Col. (1)	Column (2)	Column (4)	Column (6)	Column (8)	
01	Jammu & Kashmir	883	859	887	No
02	Himachal Pradesh	974	906	983	No
03	Punjab	893	846	899	No
04	Chandigarh	818	867	812	No
05	Uttarakhand	963	886	975	No
06	Haryana	877	830	885	No
07	NCT of Delhi	866	866	866	Zero
08	Rajasthan	926	883	935	No
09	Uttar Pradesh	908	899	910	No
10	Bihar	916	933	912	No
11	Sikkim	889	944	883	No
12	Arunachal Pradesh	920	960	913	No
13	Nagaland	931	944	929	No
14	Manipur	987	934	995	No
15	Mizoram	975	971	976	No

So, the required States/UTs are = Delhi only (Exactly equal to 50%).
Hence option (a) is the answer.

Farm Villa

Set 30

Direction for questions 1–4

Go through the data set given below and solve the questions based on it.

Year	Farm population, April		Farm employment (thousand)			Crops harvested (millions of kg) acres)	Selected indexes of input use (1992=100)						
	No. (thousands)	As % of total population	Total	Self – employed unpaid workers	Hired Workers		Total	Farm labour	Farm real estate	Durable equipment	Energy	Agricultural Chemicals	Feed, seed and purchased live stock
2000	4591	1.9	2891	2000	892	322	101	102	101	105	100	95	99
2001	4632	1.9	2877	1968	910	318	102	106	100	103	101	100	99
2002	–	–	2810	1944	866	319	100	100	100	100	100	100	100
2003	–	–	2800	1942	857	308	101	96	97	97	100	105	101
2004	–	–	2767	1925	842	321	102	96	94	94	103	106	102

Q1. The average indices of input use from 2000 to 2004 for farm labour, durable equipment and energy, in their increasing order, can be arranged as

(a) farm labour, durable equipment, energy (b) durable equipment, energy, farm labour

(c) durable equipment, farm labour, energy (d) energy, durable equipment, farm labour

Q2. If hired workers include 80% locals and 20% migrants in 2000, and 75% locals and 25% migrants in 2001, what is the percentage rise or drop in migrant hired workers from 2000 to 2001?

(a) Drop 13.25% (b) Rise 13.23% (c) Drop 27.5% (d) Rise 27.5%

Q3. What is the ratio of the number of hired workers to that of self-employed unpaid workers for the duration 2001 to 2003?

(a) 0.62:1 (b) 0.45:1 (c) 2.1:1 (d) 1.6:1

Q4. Which of the following is correct?

(I) In 2003, minimum area was harvested for crops but maximum agricultural chemicals were used.

(II) In 2004, both the number of self-employed unpaid workers and the use of durable equipments fell by minimum.

(a) I only (b) II only (c) I and II (d) None of these

Answer Grid

Q1	Q2	Q3	Q4
c	d	b	d

STORYLINE

Table presents the farm data for the years 2000 to 2004. Right-hand side of the table has the index for input use taking base year as 1992. In the base year 1992, index = 100. To solve the question set given ahead, you need not get into the technicalities of the question, be focused towards what is asked.

Solutions to Q1–Q4

A1. Average of farm labour indices $= \dfrac{102+106+100+96+96}{5} = 100$. And so on.

Similarly, average indices for other can also be calculated.
Hence option (c) is the answer.

A2. In 2000, migrant hired workers =178.5(20% of 892)
In 2001, migrant hired workers =227.5(25% of 910)

So, % rise $= \dfrac{227.5-178.4}{178.4} \times 100 = 27.5\%$

Note from Author: I would suggest you to use approximation techniques instead of actual calculation. It can be clearly seen that there is increase in the number of migrant hired workers in the year 2001 over the year 2000, hence options (a) and (c) are eliminated.

Now take down the approximated values:
In 2000, migrant hired workers =178.5(20% of 892) = 180 (Approximately)
In 2001, migrant hired workers =227.5(25% of 910) = 230 (Approximately)
So, increase = 50 (Approximate). Since our base (or denominator) is the value of 2000 = 180, hence it can be clearly seen that 50 is above 20% and less than 30% of 180. Hence we eliminate option (b) as the answer. We are not left out with option (d) only.
Hence option (d) is the answer.

A3. In 2001–2003:
Total hired workers = 910+866 + 857 = 2633 thousand
Total self-employed workers =1968 + 1944 + 1942 = 5854 thousand.

Required ratio $= \frac{2633}{5854}$

We will again use the approximation techniques to get down to the answer.

Obviously, the ratio $\frac{2633}{5854}$ is less than 1. Hence options (c) and (d) are eliminated.

Further, 50% of 5854 = 2900 (Approximately). Numerator is less than this. So ratio would be less than 0.5. There is only one option fitting the criteria – option (b). Hence option (b) is the answer.

A4. (I) is not correct because maximum agricultural chemical was used in 2003 where the index is 106. Even (II) is not correct.
Hence option (d) is the answer.

Domestic Watch Market Set 31

Direction for questions 1–4

Go through the data sets given below and solve the questions based on it.

Following data sets given below present the statistics related to the Indian watch industry.

There are five companies and their respective market share of the year 2010 is given in the pie chart given below.

Chart 1:

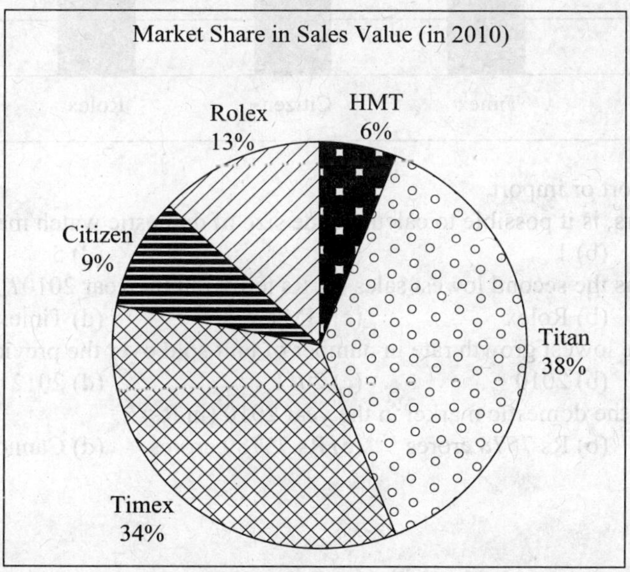

Following line chart present the Titan Sales volume (in million units) for the years 2008 to 2012. All the values are even numbers.

Chart 2:

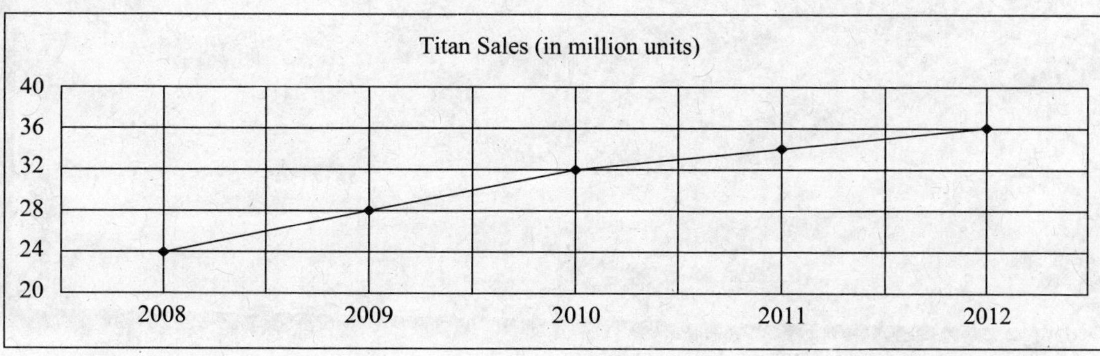

Following bar chart presents the average selling price (in Rs) of these companies in the year 2010:
Chart 3:

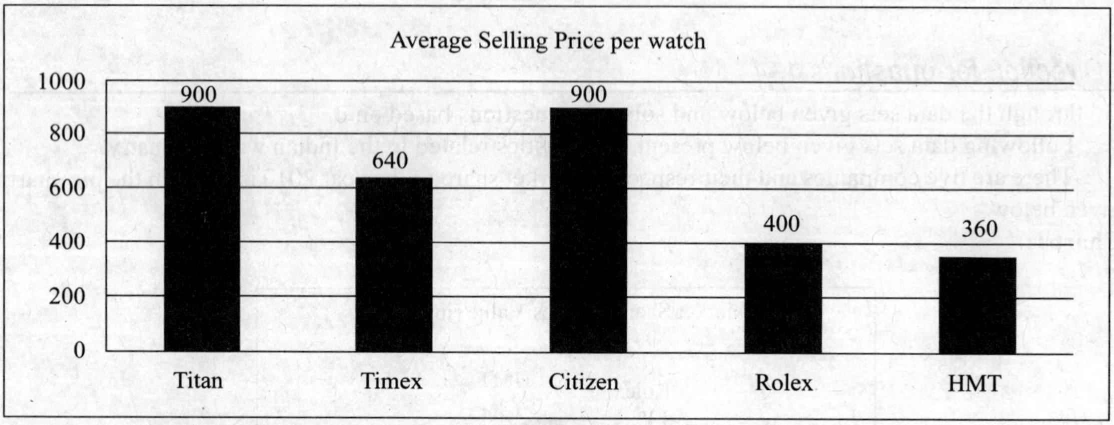

Assume there is no export or import.

Q1. For how many years, is it possible to calculate the size of domestic watch market (in Rs)?
(a) 0 (b) 1 (c) 2 (d) 5

Q2. Which company has the second lowest sales (in Rs terms) in the year 2010?
(a) Citizen (b) Rolex (c) HMT (d) Timex

Q3. Which year saw the lowest growth rate in number of unit sold over the previous year for Titan?
(a) 2009 (b) 2010 (c) 2011 (d) 2012

Q4. What is the size of the domestic market in the year 2010 (in Rs)?
(a) Rs 757 crores (b) Rs 7578 crores (c) Rs 75789 crores (d) Cannot be determined

Answer Grid

Q1	Q2	Q3	Q4
b	b	d	b

STORYLINE

Sales Value = Sales Volume × Average Price Per Unit

$$\text{Total Market Size} = \frac{\text{Sales Value of Company x}}{\text{Market Share (in percentage) of company x}} \times 100$$

Further, read the solution of Q1.

Solutions to Q1–Q4

A1. Question is asking for size of domestic market (in Rs). Out of three different data sets given in the question (a pie chart, a line chart and a bar chart), we have Rupees value in only one data set—Bar chart. In pie chart and line chart, we do not have any information in Rupees Terms. So to answer any question in Rupees terms, we need to use Chart 3—Bar chart.

How do we calculate the size of domestic watch market?

Chart 1 provides the market share in sales volume in 2010. Chart 2 provides the sales volume of Titan from 2008 to 2012.

Using chart 2, Sales volume of Titan in 2010 = 32 million units and using chart 3, average selling price = Rs 900/watch. So, total market value for Titan in 2010 = 32 million **units × Rs 900/watch = Rs X (Assume) [we are not required to calculate this value as question is ONLY asking for the possibility of market size calculation].**

Now, using chart 1, market share of Titan is known = 38% = Rs X. Using this, we can calculate 100% = Total Market Size of domestic watch market.

Hence it is possible to calculate only for the year 2010. We do not have Average Selling Price of Titan for year other than 2010. So we cannot calculate the domestic market size for any year other than 2010.

Hence option (b) is the answer.

A3. Now this is a no-brainer. You just have to calculate the percentage growth in the number of units sold of Titan.

2012 saw the minimum growth rate = 4.6%. Hence option (d) is the answer.

A4. Using the data from A1,

So, total market value for Titan in 2010 = 32 million units × Rs 900/watch = Rs 28800 million

Titan has a market share of 38% in 2010.

Hence 38% = Rs 28800 million

⇒ 100% = Rs 75789.47 million = Rs 7578.9 Crores. Hence option (b) is the answer.

Air Lines' Dilemma

Set 32

Direction for questions 1–3

Read the passage given below and solve the questions based on it.

In calendar year 2008, there was turbulence in the air as Jet Airways' Chairman pondered what course of action the airline should take. Air India was also struggling with the same dilemma. Two of India's largest airlines, Air India and Jet Airways, had sounded caution on their fiscal health due to mounting operational costs. A daily operational loss of $2 million (Rs 8.6 crore) had in fact forced Jet Airways to put its employees on alert. Jet's senior General Manager had termed the situation as grave. Jet's current losses were $2 million a day (including Jet-Lite). The current rate of Jet Airways' domestic losses was $0.5 million (Rs 2.15 crore) and that of JetLite was another $0.5 million. International business was losing over $1 million (Rs 4.30 crore) a day.

The situation was equally grave for other national carriers. Driven by mounting losses of almost Rs 10 crores a day, Air India, in its merged avatar, was considering severe cost-cutting measures like slashing employee allowances, reducing in-flight catering expenses on short-haul flights and restructuring functional arms. The airline also considered other options like cutting maintenance costs by stationing officers at hubs, instead of allowing them to travel at regular intervals. Jet Airways, Air India and other domestic airlines had reasons to gel worried, as 24 airlines across the world had gone bankrupt in the year on account of rising fuel costs. In India, operating costs had gone up 30–40%. Fuel prices had doubled in the past one year to Rs 70,000 per kilolitre, forcing airlines to increase fares. Consequently, passenger load had fallen to an average 55–60% per flight from previous year's peak of 70–75%. Other airlines faced a similar situation; some were even looking for buyers. Domestic carriers had lost about Rs 4,000 crore in 2007–08 with Air India leading the pack. "As against 27% wage bill globally, our wage bill is 22% of total input costs. Even then we are at a loss," an Air India official said. Civil aviation ministry, however, had a different lake. "Air India engineers go to Dubai every fortnight to work for 15 days and stay in five star hotels. If they are stationed there, the airline would save Rs 8 crore a year. This is just the tip of the iceberg. There are several things we can do to reduce operational inefficiency". According to analysts, Jet Airways could be looking at a combined annual loss of around Rs 3,000 crore, if there were no improvement in operational efficiencies and ATF prices. Against this backdrop, the airline had asked its employees to raise the service bar and arrest falling passenger load.

Q1. Which of the followings are the reasons for Jet Airways not doing well?

1. Rising ATF prices
2. Reduced passenger load
3. Declining service quality
4. Staff travelling to Dubai

(a) 1 and 2
(b) 2 and 3
(c) 1, 2 and 3
(d) 1, 2 and 4
(e) 1, 2, 3 and 4

Q2. The total loss for the airline industry was likely to be Rs 10,000 crore. Jet Airlines lost Rs 3,000 crore, Air India lost Rs "X" crore and "rest of the airlines" lost Rs "Y" crore. What was the loss for the "rest of the airlines", in 2008?

(a) Cannot be determined
(b) Rs 3,350 crore
(c) Rs 3,690 crore
(d) Rs 3,340 crore
(e) None of the above

Q3. Suppose fuel constitutes 30% of the revenues, do you think airlines would be in a better situation by reducing their fare?

(a) Yes
(b) Data insufficient to reach decision
(c) No
(d) It would not matter
(e) None of the above

■ Answer Grid

Q1	Q2	Q3
a	b	b

STORYLINE

This set is the combination of Reading Comprehension and Data Interpretation skills. One is supposed to find out the meaningful data within the passage mentioned.

■ Solutions to Q1–Q3

A1. Only "Rising ATF prices" and "Reduced passenger Load" have been mentioned as the reasons for Jet Airways not doing well. Airlines has asked its employees to raise the service bar—but using this statement, this cannot be assumed that service standards is low. Staff travelling to Dubai has been mentioned in case of Air India.
Hence option (a) is the answer.

A2. Total loss for the airlines industry = Rs 10,000 crores
Loss of Jet Airlines = Rs 3,000
Loss to Air India = Rs 10 crores × 365 days = Rs 3650 crores (Air India's daily loss is mentioned in the second line of 2nd para).
Hence total loss to all the other airlines = (Rs 10,000 – Rs 3000 – Rs 3650) crores = Rs 3350 crores. Hence option (b) is the answer.

A3. Nothing can be said for sure regarding the revenue of Airlines. Reducing the fare may improve the passenger load, but nothing can be ascertained if this will increase the net revenue or not?
Hence option (b) is the answer.
Option (d) cannot be the answer because "It matter to reduce the fare".

Mode of Powering Railway Engines

Set 33

Direction for questions 1–3

Go through the data set given below and solve the questions based on it.

In the five countries listed below, information with regard to mode of powering railway engines is provided. Railway engines can be powered only by one of the three modes—Natural Gas, Diesel or Electricity. A single railway engine will use only one of the three modes mentioned for whole of its life. Assume that all the data and questions given below pertain to the same time period.

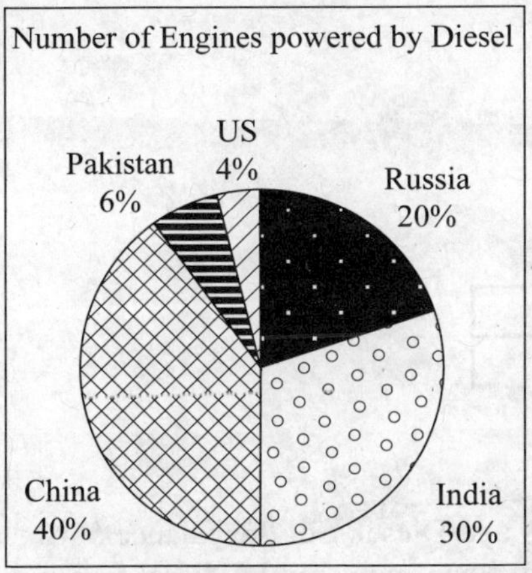

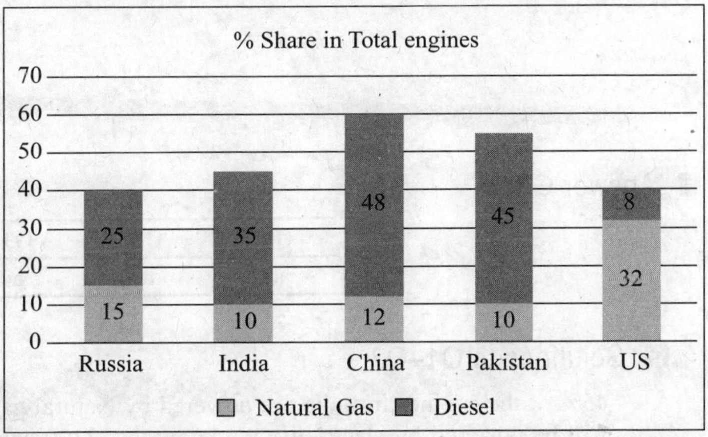

Also given that the total number of engines powered by diesel in the given five countries above = 7030 (as given in the pie chart)

Q1. What is the number of engines in India that are powered by electricity?

(a) 3300 (b) 2700 (c) 6000 (d) Cannot be determined

Q2. If the number of engines in Russia accounts for 5% of the total number of engines in the world, then what percentage of the total engines in the world are powered by diesel?

(a) 110,000 (b) 27,500 (c) 14100 (d) Cannot be determined

Q3. Amongst the five countries, what is the average percentage of engines that are powered by electricity?

(a) 52.4% (b) 47.6% (c) 55.8% (d) 44.2%

Rough Work

Answer Grid

Q1	Q2	Q3
a	d	a

Solutions to Q1–Q3

A1. 45% of the engines in India are powered by Natural gas and Diesel. So, the balance 55% will be powered by electricity. We know that 2100 engines are powered by diesel which accounts for 35% of the total engines. Therefore, 55% of the total engines = $\frac{2100}{35} \times 55 =$ 3300. Hence option (a) is the answer.

A2. We can find out the total number of engines in the world with the information available. But we do not know the makeup of the engines of the world in terms of their being powered by different modes. Hence cannot be determined.
Hence option (d) is the answer.

A3. Average % of engines powered by electricity = (Number of engines powered by electricity/ Total number of engines) × 100

$= \frac{11340000}{21625} \times 100 = 52.4\%$. Hence option (a) is the answer.

Missing Data in Population Density

Set 34

Direction for questions 1–4

Go through the data set given below and solve the questions based on it.

Following bar chart and table provides the Census Data for India during the given period.

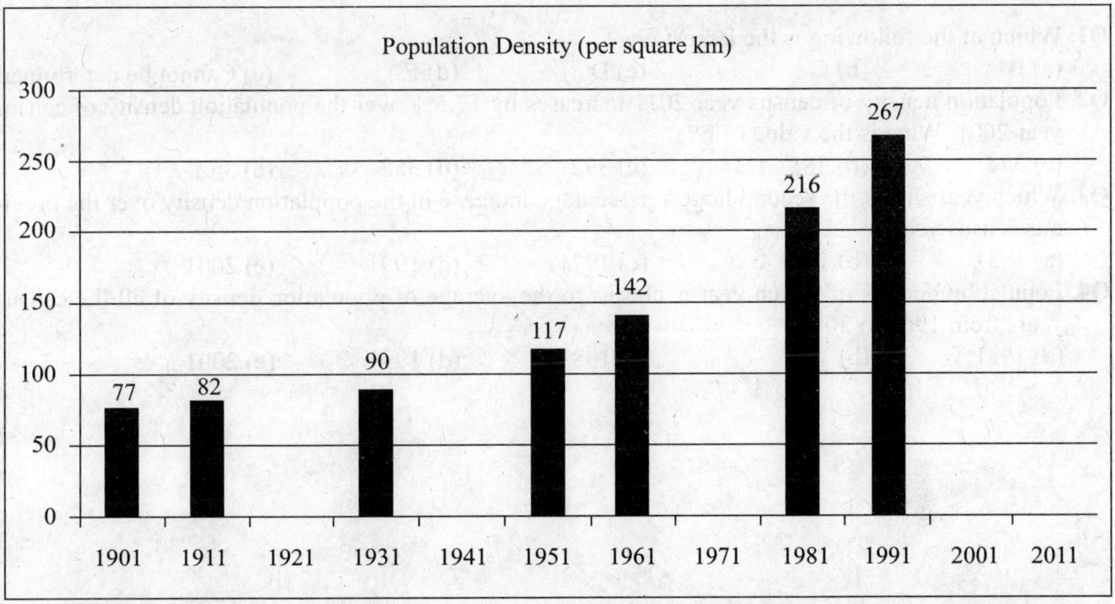

As it can be seen that in the bar chart, population density data for few years namely 1921, 1941, 1971, 2001 and 2011 have been omitted due to some technical fault. Assume these values to be equal to A, B, C, D and E respectively.

Following table provides the absolute increase in the density of the census year with respect to the previous census year.

Census year	Absolute increase from previous census
1901	NA
1911	5
1921	–1
1931	9
1941	13

Census year	Absolute increase from previous census
1951	14
1961	25
1971	35
1981	39
1991	51
2001	58
2011	Not mentioned

Q1. Which of the following is the largest one?
(a) B (b) C (c) D (d) E (e) Cannot be determined

Q2. Population density of census year 2011 increases by 17.5% over the population density of census year 2001. What is the value of E?
(a) 372 (b) 382 (c) 392 (d) 352 (e) 362

Q3. Which year shows the second largest percentage increase in the population density over the previous census year?
(a) 1931 (b) 1941 (c) 1971 (d) 1991 (e) 2001

Q4. Population density of which year is closest to the average of population density of all the census years from 1901 to 2001?
(a) 1931 (b) 1961 (c) 1981 (d) 1991 (e) 2001

Rough Work

Answer Grid

Q1	Q2	Q3	Q4
e	b	d	b

STORYLINE

Barchart provides the population density of the census years – though data for some of the years is missing. To find out the missing values in the bar chart, table provides the absolute increase in the census year with respect to the previous census year. For example, census year of 1941 = 13 more from the population density of 1931 = 90 + 13 = 103.

Solutions to Q1–Q4

A1. It can be seen that density has increased in all the census years over the previous year except (a) in 1921 (b) in 2011 – we do not have information. Hence population density for 2011 = E is not known to us.
So final answer is—cannot be determined. Hence option (e) is the answer.

A2. Given that Population density in 1991 = 267
So population density of 2001 = 267 + 58 = 325. To calculate the population density of 2011, we have been given the clue in terms of percentage increase over 2001 population density.
Population density of 2011 = Population density of 2001 + increase of 17.5% over this = 325 + 17.5% of 325

$$17.5\% = 20\% - 2.5\% = \frac{1}{5} - \frac{1}{40}$$

$$\frac{1}{5} \text{ of } 325 = 65 \text{ and } \frac{1}{40} \text{ of } 325 = 8 \text{ (approximately)}$$

So, 17.5% = 65 – 8 = 57
Hence population density of 2011 = 325 + 57 = 382. Hence option (b) is the answer.

A3. Following table gives the population density and percentage change:

Census year	Density (per km^2)	% age change
1901	77	
1911	82	6.5
1921	81	–1.2
1931	90	11.1
1941	103	14.4
1951	117	13.6
1961	142	21.4
1971	177	24.6
1981	216	22
1991	267	23.6
2001	325	21.7

Second largest percentage increase occurs in 1991. Hence option (d) is the answer.

A4.

Method 1: Using approximation

Since the population density is constantly on a rise, average would be accounted for by the years that lie midway—1951 or 1961 or 1971. Census year 1951 can be easily ruled out as population density in this year is too low. So either census year 1961 or 1971 will be the answer. Since 1971 is not in the options, hence 1961 is the answer.

Method 2: Using the actual calculation

Using the table drawn to solve Q3, average population density of all the census years from 1901 to 2001 = 152. This is closest to the population density of the census year 1961. Hence option (b) is the answer.

Indian Cricket Team's Batting in ODI Series

Set 35

Direction for questions 1–4

Go through the chart given below and solve the questions based on it.

In the game of cricket, there are 11 batsmen who bat during the match of 50 overs. 1st two batsmen come to bat first and when any one of the two gets out, next batsman come to bat alongwith the one of the two batsmen left out. And so on the next batsman comes to bat—in a fixed sequence for all the three matches. Only one batsman can get out for one ball. One over comprises of 6 balls—to be thrown one by one. Though if all the batsmen get out before scheduled 50 Overs, match ends at that point (known as Overs Actually Played).

Following bar chart shows total scheduled overs and overs actually played for first match (denoted by 1), second match (denoted by 2) and third match (denoted by 3).

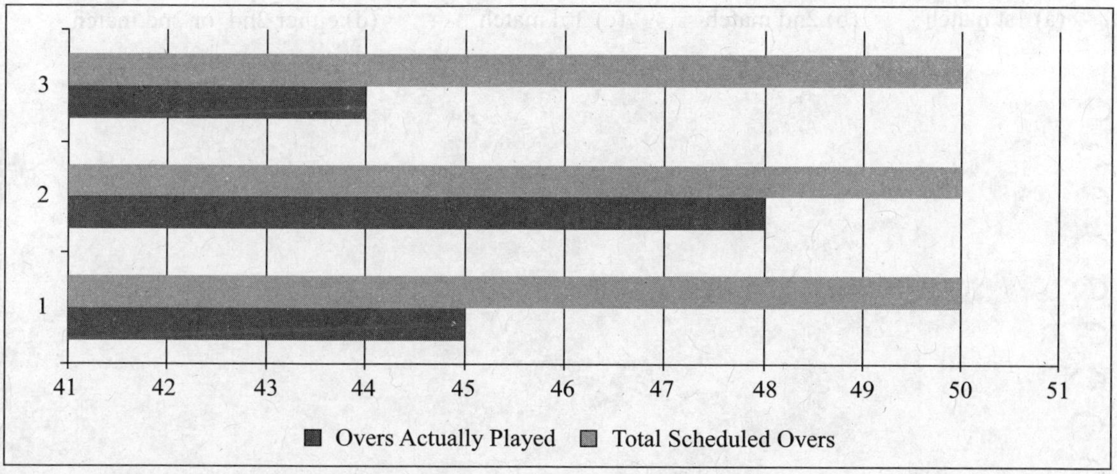

Following table shows the runs scored in three matches series. Total runs scored = Runs scored by 1st five players + Runs scored by remaining last six players + Extra runs.

India Vs Pakistan ODI Series – Runs Scored			
	1st Match	2nd Match	3rd Match
Runs Scored by 1st Five Players	18	75	80
Runs Scored by Remaining Last Six Players	187	72	70
Extra Runs	22	18	17

Q1. In which of the matches, Indian Team's average runs per over is maximum?
(a) First match (b) Second match (c) Third match (d) Either first or second match

Q2. What would have the total runs scored in the 3rd match if Indian team would have played 50 Overs with the same average which it maintained till the Overs actually played?
(a) 180 (b) 190 (c) 200 (d) 170

Q3. Mr Phani Ram, the team selector, has made following observations to improve the performance of the team:

(i) Performance of 1st five players have improved, whereas performance of last six players have gone down.

(ii) Had there been no extra runs, runs in 2nd match would have decreased by maximum percentage.

(iii) If team would have played for full 50 overs, it would have scored maximum in the first match.

How many of these observations are true?
(a) Only (i) and (ii) but not (iii)
(b) Only (ii) and (iii) but not (i)
(c) Only (i) and (iii) but not (ii)
(d) All three

Q4. If extra runs are not considered, which match has the lowest Average runs per over?
(a) 1st match (b) 2nd match (c) 3rd match (d) either 2nd or 2nd match

Rough Work

Answer Grid

Q1	Q2	Q3	Q4
a	b	c	b

STORYLINE

Question provides two sets of data—Overs and Runs Scored. While total scheduled overs is fixed at 50, overs actually played vary for all the three matches. Runs scored have been given in three parts—Runs scored by 1st five players, runs scored by remaining last six players and extra runs. Summation of these three components gives total runs. Further, Average runs per over has been defined.

Solutions to Q1–Q4

A1. Clearly it can be seen that runs scored in the 1st match is much more than the total runs scored in either 2nd match or 3rd match. Though number of overs actually played is almost similar in all the three matches – 45 overs in 1st match, 48 overs in 2nd match and 44 overs in 3rd match.
Since numerator (Total runs scored) is maximum in case of 1st match whereas denominator is almost similar, Average runs per over will be maximum for 1st match.
Hence option (a) is the answer.

A2. Total number of overs actually played in 3rd match = 44 overs
Total number of runs scored in 3rd match = 80 + 70 + 17 = 167 runs
Method 1 – Through actual calculation method.

Average runs per over in 3rd match $= \frac{167}{44} = 3.8$ runs per over

Total runs in 50 Overs = 50 × 3.8 = 190 runs
Method 2 – Through approximation method.

44 overs $\xrightarrow{\text{Percentage Increase of 12\%–13\%}}$ 50 overs

Hence runs scored will also increase by 12%–13%. Now we will not go through the actual calculation. A quick look at the options shows that options are not close, so we can have a good approximation.
Instead of increasing 167 runs (scored in 44 overs) by 12%–13%, we increase it by ONLY 10% to find a quick range of our answer.

167 $\xrightarrow{\text{Percentage Increase of 10\%}}$ 167 + 16.7 ≡ 184. Our actual answer will be a bit more than this.
Option (b) is the closest one.
Hence option (b) is the answer.

A3. Evaluating the observations one by one:
Observation (i)– Understand that you don't have to use your "Cricket" knowledge to answer this question. Taking the data on its face value, it can be clearly seen that performance of 1st five players has improved whereas performance of last six players has gone down. Hence observation (i) is correct.
Observation (ii)– This is a close call. So let us go through the actual calculation.

Percentage decrease in case of Match 1 $= \frac{22}{227} \times 100 = 9.69\%$

Percentage decrease in case of Match 2 $= \frac{18}{165} \times 100 = 10.9\%$

Percentage decrease in case of Match 3 = $\frac{17}{167} \times 100 = 10.17\%$

So, maximum percentage decrease occurs in case of Match 3. Hence Observation ii is wrong.

Observation (iii)– Since the average runs per over is maximum in Match 1 (using answer to Q1), hence runs scored in 50 overs would have been maximum in match 1.So observation iii is true.

So, only observations (i) and (iii) are correct.

Hence option (c) is the answer.

A4. This is simple calculation – similar to A1.

Option (a) can be easily eliminated as this has the maximum runs scored. Though to decide between match 2 and match 3, we are required to calculate the actual value.

average runs per over for match 2 = 3.06 runs per over and average runs per over for match 3 = 3.40 runs per over. Hence match 2 is the answer.

Hence option (b) is the answer.

Data from Forest

Set **36**

Direction for questions 1–4

Go through the data set given below and solve the questions based on it.

Below given pie charts show the percentage population of wild animals and the percentage of land use under forest cover in different regions of the world in the year 2007.

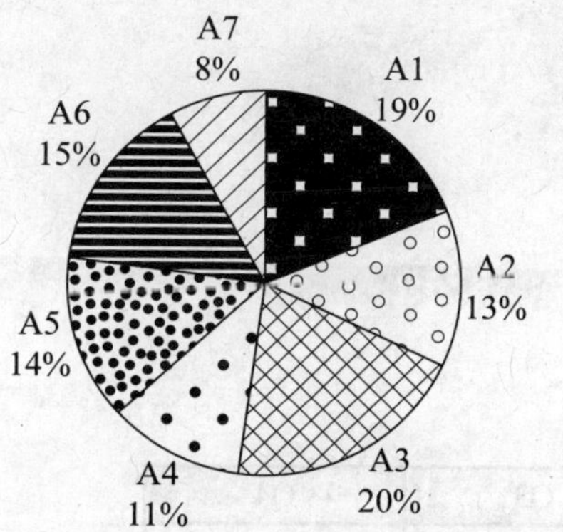

Percentage wise distribution of population of wild animals.

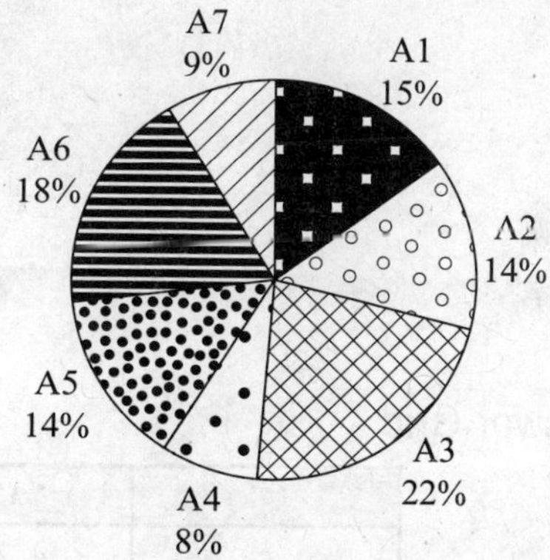

Percentage wise distribution of land under forest cover

Following are the geographical regions:

A1 - Asia, China, India, A2 – Australia, A3 – Africa, A4 – Europe, A5 – Russia, A6 - North & South America, A7 – Antarctica. Population density is obtained by dividing population by land.

$$\text{Population density} = \frac{\text{Percentage Number of Animals}}{\text{Percentage Land Area Under Forest Cover}}$$

Q1. If the population density of wild animals in Russia is 1, then what is the combined population density of (Africa and North & South America)?

(a) 0.79 (b) 0.84 (c) 0.88 (d) 0.94

Q2. Which among the following has the fourth highest population density of wild animals?

(a) A5 (b) A7 (c) A3 (d) A2

Q3. Population density of wild animals in Europe is how many times that of the population density of wild animals in Russia?

(a) 1.375 (b) 1.49 (c) 1.57 (d) 1.64

Q4. If in the next 10 years, the population of wild animals in Antarctica will increase by 2.5% more than that in the rest of the world, what will be the percentage population of wild animals in Antarctica after 10 years?

(a) 9.5% (b) 11.6% (c) 15.7% (d) Cannot be determined

Rough Work

Answer Grid

Q1	Q2	Q3	Q4
c	d	a	d

STORYLINE

Pie charts given in the question present two different data for seven regions namely A1 to A7. Left hand piechart presents the population distribution of wild animals in seven regions whereas right hand piechart presents the land under forest cover distribution in seven regions.

$$\text{Population density} = \frac{\text{Percentage Number of Animals}}{\text{Percentage Land Area Under Forest Cover}}$$

Solutions to Q1–Q4

A1. This is a straightforward question. We need to add the percentage of animals and divide it by summation of percentage of land area under forest cover.

$$\text{Required population density} = \frac{20+15}{22+18} = 0.88.$$

Hence, option (c) is the answer.

A2. If the question would have asked us ‘maximum’ or ‘minimum’, we could have solved it through process of elimination. Since question asks us to evaluate “fourth highest population density”, it is better that we calculate the density for each of the regions and identify the fourth highest.

The population densities of the areas are:

A1 = 1.266 A2 = 0.928 A3 = 0.909 A4 = 1.375

A5 = 1 A6 = 0.833 A7 = 0.89

Alternatively, we can use some guess estimation skills to shorten the calculation process: Three regions namely A1, A4 and A5 are clearly greater than or equal to 1. So, fourth highest is a value closest AND lower than 1.

Hence, option (d) is the answer.

A3. This is simple calculation.
Population density of Russia = 1 while population density of Europe = 1.375.
Hence, option (a) is the answer.

A4. As the increase in the population of the rest of the world is not known, we cannot determine the increase in Antarctica's wild life population. Hence, option (d) is the answer.

Pipeline Transportation **Set 37**

Direction for questions 1–3

Read the passage given below and solve the questions based on it.

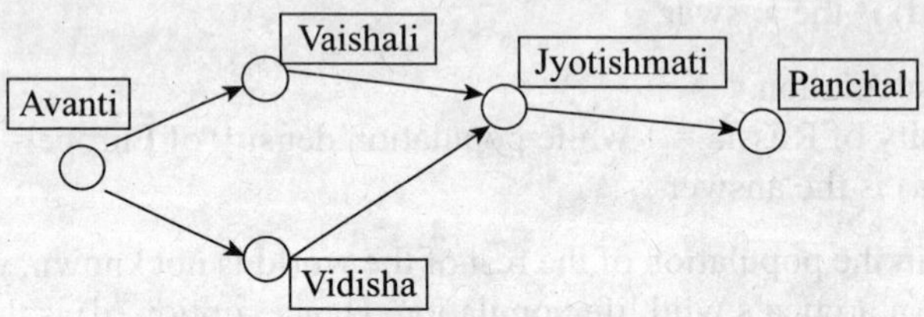

The figure shows an Oil Pipeline Network among 5 cities. Each pipeline has the load carrying capacity of 1000 and each destination has a specific requirement that has to be met. The city of Vaishali and Jyotishmati have a requirement of 400 each. The requirement at Panchal is 700 where as that at Vidisha is 200. The load can be carried in the indicated directions only. It is further given that the load carried from Vaishali to Jyotishmati is 300.

Q1. What is the free capacity in the Avanti–Vidisha pipeline?

(a) 300 (b) 200 (c) 100 (d) 0 (e) None of these

Q2. What is the free capacity in the Avanti–Vaishali pipeline?

(a) 0 (b) 100 (c) 200 (d) 300 (e) None of these

Q3. What is the quantity moved from Avanti to Vidisha?

(a) 200 (b) 800 (c) 700 (d) 1000 (e) None of these

Rough Work

Answer Grid

Q1	Q2	Q3
d	d	d

Solutions to Q1–Q3

Following is the requirement at different cities:

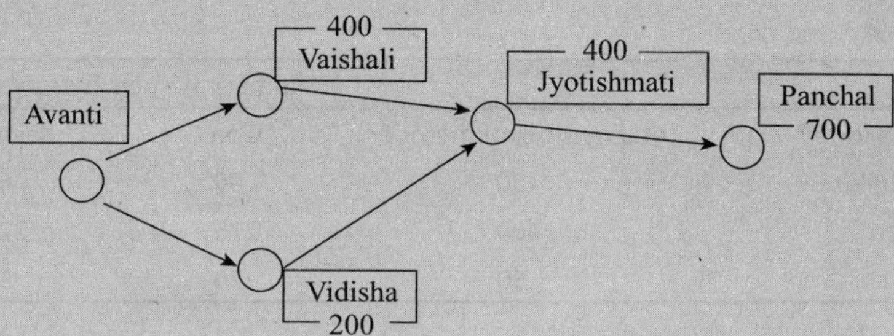

Total requirement of Jyotishmati + Panchal = 700 + 400 = 1100. Out of this, 300 comes through Vaishali–Jyotishmati pipeline. So remaining 800 shall come through Vidisha–Jyotishmati pipeline. Now, requirement at Vidisha = 200. Hence, Avanti Vidisha pipeline shall carry 200 (Requirement at Vidisha) + 800 (to be sent further to Jyotishmati) = 1000

So this is the final situation:

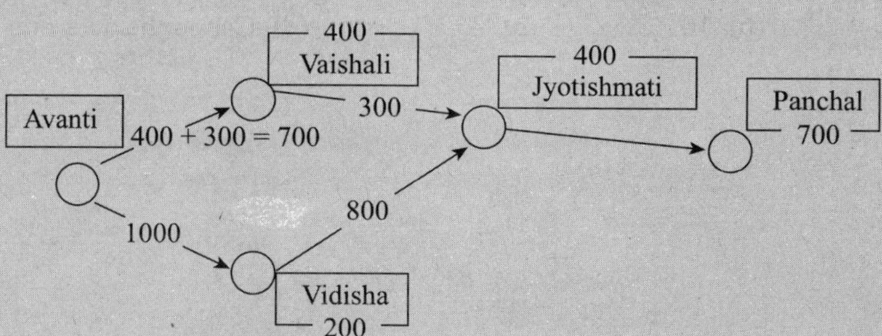

Now all the questions can be answered.

A1. Each pipeline can carry a maximum of 1000. Since Avanti–Vidisha pipeline carries 1000, free capacity in Avanti–Vidisha is zero.

A2. Total free capacity at Avanti–Vaishali pipeline is 300. Capacity of each pipeline is 1000 and demand at Vidisha is 400 and 300 flows to Jyotishamti.
So, free capacity = {1000 – (400 + 300)}= 300

A3. Quantity flowed from Avanti to Vidisha = 200 + 100 + 700 = 1000

Family Vacation

Set 38

Direction for questions 1–3

Go through the information given below and solve the questions based on it.

Following is known about the members of Thumbs-up holidays who might have gone to Goa or Kashmir or both.

		Number of members who went to	
Number of children	Total number of members	Goa	Kashmir
Less than 3 children	150	80	60
3 to 5 children	400	275	225
More than 5 children	50	20	30

Q1. What is the least number of members who have 3 to 5 children and have gone to at least one of the destinations is:

(a) 225 (b) 275 (c) 100 (d) Cannot be determined

Q2. Minimum number of members for all the categories and have gone to at least one of the destinations is:

(a) 385 (b) 375 (c) 305 (d) 315

Q3. How many of the members have definitely not gone to any of the two destinations?

(a) 0 (b) 10 (c) 25 (d) Cannot be determined

Rough Work

Answer Grid

Q1	Q2	Q3
b	a	b

Solutions to Q1–Q3

A1. In the category 3 to 5 children, there are 275 members who have gone to Goa. We are not sure about the 225 members who have gone to Kashmir that if they are different than the members who have gone to Goa. So for these 225 families going to Kashmir, we are not sure about the extent of overlapping with the families going to Goa. It might be a possibility that these 225 members who have gone to Kashmir have also gone to Goa. Hence minimum number of members who have gone to atleast one of the destinations = 275.
Hence option (b) is the answer.

A2. Taking a cue from solution to Q1, following numbers are to be added:

Number of members who went to	
Goa	Kashmir
80	60
275	225
20	30

Answer to this question would be = 80 + 275 + 30 − 385. Hence option (a) is the answer.

A3. We are not sure of the members in the category '3 to 5 children' or 'more than 5 children' that any of these members have NOT gone to any of the destinations. Although we are sure of the fact that in the category 'Less than 3 children', there are at least 150 – (80+60) = 10 members who have not gone to any of the destinations. Hence option (b) is the answer.

Examination Report Card Set 39

Direction for questions 1–4

Refer to the table given below and answer the questions that follow.

Following table gives some incomplete information on the marks obtained by four persons Amit, Binit, Charu and Dilip in four different subjects.

	Physics	Chemistry	Maths	English	Total
Amit	80	70	-	80	-
Binit	60	-	60	-	-
Charu	-	-	60	-	220
Dilip	90	-	70	60	-

Additional information:

Amit's average marks was 72.5, in all the four subjects.

The average marks in physics for all four persons was 67.5.

Binit's average marks for all four subjects was 1.5 times Charu's marks in physics.

Binit and Dilip scored the same marks in English. The average English marks were the same as the average Physics marks.

The total marks in Chemistry was 10 less than the total marks in physics for all four persons.

Q1. Who scored the lowest total?

(a) Amit (b) Binit (c) Charu (d) Dilip

Q2. The average marks for all four persons (for all the four subjects) is (approx.)

(a) 65.625 (b) 81.225 (c) 99.5 (d) 105.6

Q3. Who is the highest scorer in chemistry?

(a) Amit (b) Binit (c) Charu (d) Dilip

Q4. Who is the lowest scorer in physics?

(a) Amit (b) Binit (c) Charu (d) Dilip

Rough Work

Answer Grid

Q1	Q2	Q3	Q4
c	a	d	c

Solutions to Q1–Q4

From the given information, the following table can be prepared:

	Physics	Chemistry	Maths	English	Total
Amit	80	70	60	80	290
Binit	60	60	60	60	240
Charu	40	50	60	70	220
Dilip	90	80	70	60	300

A.1 (c) A.2 (a) A.3 (d) A.4 (c)

Returns from Stocks Set 40

Direction for question 1–4

Go through the data set given below and solve the questions based on it.

Name of Stock	Date of Auction	Date of Issue	Notified Amount	Devolvement on PDs	Cut off price (in Rs)	Cut off yield (%)	Date of Maturity
7.56% GS 2014	02-Apr-09	06-Apr-09	8000	0	103.47	6.80	03-Nov-14
7.94% GS 2021	02-Apr-09	06-Apr-09	4000	567	102.5	7.62	24-May-21
6.05% GS 2019	09-Apr-09	13-Apr-09	8000	0	95.05	6.75	02-Feb-19
7.50% GS 2034	09-Apr-09	13-Apr-09	4000	0	97.35	7.74	10-Aug-34
7.56% GS 2014	17-Apr-09	20-Apr-09	8000	0	106.75	6.10	03-Nov-14
8.24% GS 2027	17-Apr-09	20-Apr-09	4000	0	107.82	7.44	15-Feb-27
6.05% GS 2019	24-Apr-09	27-Apr-09	8000	0	99.38	6.13	02-Feb-19

Q1. For how many stocks, date of maturity is minimum 5 years ahead of date of auction?
(a) All but 1 stock (b) All but 2 stocks
(c) All but 3 stocks (d) None of these

Q2. Nomenclature of the stocks is done in the following way:
7.56% GS 2014 - It means that desired yield from the Government stock (GS) in 2014 is 7.56%.
For how many of the stocks desired yield is more than cut-off yield?
(a) 3 (b) 4 (c) 5 (d) 2

Q3. How many stocks with a cut-off price of more than Rs 100 have their cut-off yield more than its desired yield (use definition from above question if required)?
(a) 0 (b) 1 (c) 2 (d) 3

Q4. Mr Robert is quite an ambitious person. He wants to obtain a minimum desired yield of 8%. For that to happen, he buys exactly two stocks in different ratio. How many combinations can he have, out of 7 stocks given, such that his minimum desired yield is atleast 8% (use information from previous question)?
(a) Not possible (b) 6 (c) 7 (d) None of these

Rough Work

Answer Grid

Q1	Q2	Q3	Q4
d	b	a	b

STORYLINE

Best way to go through this set of questions is to take the questions on its face value without understanding the meanings attached to the terms (though they are self descriptive in nature, and one is not required to scratch one's head to make meaning out of it).

Solutions to Q1–Q4

A1. The least occurs in case of 7.56% GS 2014 which is more than 5 years. Hence answer is – all the years. Hence answer is option (d) None of these.

A2. For 4 stocks, the desired yield is more than cut off yield. Hence option (b) is the answer.

A3. Following stocks are having their cut-off price more than Rs 100:

Name of Stock	Cut off price
7.56% GS 2014	103.47
7.94% GS 2021	102.5
7.56% GS 2014	106.75
8.24% GS 2027	107.82

None of these stocks have their cut-off yield more than its desired yield. Hence option (a) is the answer.

A4. To obtain a minimum desired yield of 8%, Mr Robert has to take 8.24% GS, as for remaining all the stocks, minimum desired yield is 8%. (One cannot have above 8% yield by mixing two stocks with less than 7% stock. It means that one stock has to be higher than 8 % and other has to be less than 8%).
He can mix 8.24% GS with any one of the remaining 6 stocks to obtain minimum 8% desired yield.
Hence option (b) is the answer.

Gold Cup Tournament Set 41

Direction for questions 1–5

Read the passage given below and solve the questions based on it.

Sixteen teams have been invited to participate in the ABC Gold Cup cricket tournament. The tournament is conducted in two stages. In the first stage, the teams are divided into two groups. Each group consists of eight teams, with each team playing every other team in its group exactly once. At the end of the first stage, the top four teams from each group advance to the second stage while the rest are eliminated. The second stage comprises of Elimination Round matches in several rounds. A round involves one match for each team. The winner of a match in a round advances to the next round, while the loser is eliminated. The team that remains undefeated in the second stage is declared the winner and claims the Gold Cup.

The tournament rules are such that each match results in a winner and a loser with no possibility of a tie. In the first stage, a team earns one point for each win and no points for a loss. At the end of the first stage, teams in each group are ranked on the basis of total points to determine the qualifiers advancing to the next stage. Ties are resolved by a series of complex tie-breaking rules so that exactly four teams from each group advance to the next stage.

Q1. What is the total number of matches played in the tournament?

(a) 28 (b) 55 (c) 63 (d) 35 (e) None of these

Q2. The minimum number of wins needed for team in the first stage to guarantee its advancement to the next stage is

(a) 5 (b) 6 (c) 7 (d) 4 (e) None of these

Q3. What is the highest number of wins for a team in the first stage in spite of which it would be eliminated at the end of first stage?

(a) 1 (b) 2 (c) 3 (d) 4 (e) None of these

Q4. What is the number of rounds in the second stage of the tournament?

(a) 1 (b) 2 (c) 3 (d) 4 (e) None of these

Q5. Which of the following statements is true?

(a) The winner will have more wins than any other team in the tournament.

(b) At the end of the first stage, no team eliminated from the tournament will have more wins than any of the teams qualifying for the second stage.

(c) It is possible that the winner will have the same number of wins in the entire tournament as a team eliminated at the end of the first stage.

(d) The number of teams with exactly one win in the second stage of the tournament is 4.

(e) None of these

Rough Work

Answer Grid

Q1	Q2	Q3	Q4	Q5
c	b	e	c	c

STORYLINE

There are two rounds to decide the winner—First round is Round Robin League Tournament, where 8 teams are partitioned in two groups, where each team plays one match with other team. Second round is Elimination tournament.

Total number of matches in ONE Round in Round Robin League Tournament

(with n teams) $= \dfrac{n(n-1)}{2}$

Total number of matches in Elimination Tournament with n teams = (n – 1)

Solutions to Q1–Q5

A1. There shall be 8 teams in each group. Each team in a group shall be playing with every other team. Hence, total number of matches shall be $\dfrac{(7 \times 8)}{2} = 28$ in one group. Hence, in both the groups, there shall be 56 matches. This is for the first stage. Thereafter, there are 8 teams in knockout rounds from which one winner emerges, or 7 losers are identified. Hence, 7 more matches, i.e. in all 63 matches. Hence option (c) is the answer.

A2. Under extreme conditions, the following can be the maximum number of matches won.

Teams	A	B	C	D	E	F	G	H
A	NA	Ö	Ö	Ö	Ö	Ö	×	×
B	×	NA	Ö	Ö	Ö	Ö	Ö	×
C	×	×	NA	Ö	Ö	Ö	Ö	Ö
D	×	×	×	NA				
E	×	×	×		NA			
F	×	×	×			NA		
G	Ö	×	×	Ö	Ö	Ö	NA	Ö
H	Ö	Ö	×	Ö	Ö	Ö	×	NA

In the above table,
Ö represents win, × represents loss. (A match between A and A is not possible since single team cannot play with itself). From the table, A, B, C, G and H can win five matches each. Hence, we cannot decide that which team will qualify for the second round. Therefore, the minimum number of wins that can assure a place in the second stage is 6.
Hence option (b) is the answer.

A3. Looking at the solution of Q2, even a team wins 5 matches, it can be eliminated. Hence option (e) is the answer.

A4. Total number of teams qualifying for the 2nd round = 8 teams.

Rounds	Total number of teams playing in this round	Total number of matches	Total number of Teams going to next Round
Round 1	8	4	4
Round 2	4	2	2
Round 3	2	2	1

Hence option (c) is the answer.

A5. Let us evaluate all the options one by one:
Option (a) - The winner will have more wins than any other team in the tournament
Explanation – Consider the situation where a team 'N' in Pool A has won 7 matches in 1st round and gets promoted to 2nd round, wins its 1st two matches in the 2nd round. Consider another team 'M' in pool B has won ONLY 5 matches in 1st round and wins all its matches in the 2nd round. Even the eventual match between team M and team N is won by team M. Finally, team M will win the tournament.
Total matches won by team N = 9
Total matches won by team M = 8
Hence option (a) is not definitely right.
Option (b) – Explanation given to discard option (a) holds true even for option (b). Hence option (b) is not necessarily true.
Option (c) – This is possible. Hence option (c) is the answer.

FDI

Set 42

Direction for questions 1–4

Go through the data set given below and solve the questions based on it.

Following table gives the top 7 sectors attracting FDI equity inflow during the period 2001–02 to 2010–11 (as given in Cumulative Inflows which is the summation of all the equity inflows during the period 2001–02 to 2010–11 in that sector).

Sectors attracting maximum FDI Equity Inflows (Rs Crores)					
Rank	Sector	2008–09	2009–10	2010–11	Cumulative Inflows
1	SERVICES SECTOR	28,516	20,776	15,539	120,771
2	COMPUTER	7,329	4,351	3,571	48,400
3	TELECOM	11,727	12,338	7,546	48,220
4	REAL ESTATE	12,621	13,586	5,149	43,192
5	CONSTRUCTION	8,792	13,516	5,077	40,770
6	AUTOMOBILES	5,212	5,754	6,008	26,831
7	POWER	4,382	6,908	5,709	26,712

Q1. Which sector has attracted third highest FDI equity inflows from the year 2001–02 to 2007–08?
(a) Computer (b) Telecom (c) Real Estate (d) Power

Q2. Telecom sector accounts for 8% of total cumulative inflows during the whole period. What percentage of total cumulative inflows is accounted for by Services sector during the whole period?
(a) 23% (b) 20% (c) 18% (d) Cannot be determined

Q3. Out of the mentioned sector, how many sectors have seen a consistent rise or consistent decline across the years from 2008–09 to 2010–11?
(a) 2 (b) 3 (c) 4 (d) 1

Q4. For the year 2008–09, which sector accounts for the maximum percentage of the total FDI equity inflows of that particular sector for the whole period 2001–02 to 2010–11?
(a) Computer (b) Telecom (c) Real Estate (d) Services Sector

Rough Work

Answer Grid

Q1	Q2	Q3	Q4
b	b	b	c

Solutions to Q1–Q4

A1. To solve this question, we need to add the FDI equity inflows from 2008–09 to 2010–11 and subtract this from cumulative inflows. We are not required to do the actual calculation. We will just take the Thousands value of all the years. We can also observe that answer has to be either Computer or Telecom (after doing preliminary observation of data).

	2008–11	Remaining Years
SERVICES SECTOR	63	57
COMPUTER	14	32
TELECOM	30	17
REAL ESTATE	30	11
CONSTRUCTION	26	13
AUTOMOBILES	16	10
POWER	16	10

Third highest is for Telecom. Hence option (b) is the answer.

A2. This is a simple calculation based question.
Given that 8% = 48,220 $\Rightarrow$ 1% = 6000 (Approx.)
So, 120,771 = 20%
Hence (b) is the answer.

A3. Following are the sectors:
SERVICES SECTOR, COMPUTER, AUTOMOBILES
Hence option (b) is the answer.

A4. We have to choose one of these four options. Since options do not mention "none of these", we will focus upon these four options only.

Percentage to total FDI Equity in case of Computer = $\frac{285}{1207}$

Percentage to total FDI Equity in case of Telecom = $\frac{11727}{48220}$

Percentage to total FDI Equity in case of Real Estate = $\frac{12621}{43192}$

Percentage to total FDI Equity in case of Services Sector = $\frac{28516}{120771}$

Let us first go through elimination:

In case of three options, namely, Computer, Telecom, Real Estate, Base (Denominator i.e., Cumulative Inflows) is approximately, same. Numerator of Real Estate is maximum, hence out of these three options, Real estate is maximum.

So now we have to compare Real Estate with Services Sector:

Real Estate $\frac{12621}{43192}$ $\frac{28516}{120771}$ Services Sector

First of all, let us eliminate last two digits from the numerator and denominator from both the sectors:

Real Estate $\frac{126}{431}$ $\frac{285}{1207}$ Services Sector

It can be seen that Numerator of Services sector is approximately 2.25 times of the numerator of Real Estate, but denominator of Services sector is more than 2.25 times of denominator of Real Estate (we just want to know if it is more than 2.25 or less than 2.25, we don't need the exact value).

2.25 times

Real Estate $\frac{126}{431}$ $\frac{285}{1207}$ Services Sector

>2.25 times

Hence Real Estate is the answer. Hence option (c) is the answer.

Inventory Management Set 43

Direction for questions 1–4

Go through the data set given below and solve the questions based on it.

CP of One Television for Jamshedpur Electronics	Rs 1150/unit
CP of One IPod for Jamshedpur Electronics	Rs 1000/unit

Area/Month		January	February	March
Sales in Bistupur				
	Television	900	1050	1200
	Ipods	15750	16800	17850
Sales in Sakchi				
	Television	1800	2100	2400
	Ipods	9450	10080	10710
Sales in Kadma				
	Television	6300	7350	8400
	Ipods	6300	6720	7140

Q1. In a period from January to march, Jamshedpur Electronics sold 3150 units of Television, having started with a beginning inventory of 2520 units and ending with an inventory of 2880. What was the value of order placed (Rupees in thousands) by Jamshedpur Electronics during the three months period?

(a) Rs 280800
(b) Rs 4036500
(c) Rs 2232023
(d) Rs 2520023
(e) Rs 2808080

Q2. What was the total value of surcharge paid – at the rate of 14% of sales value – by Jamshedpur Electronics, in the month of January on Television?

(a) Rs 1852202
(b) Rs 1854800
(c) Rs 1842531
(d) Rs 1449000
(e) Cannot be determined

Q3. 10% of sales price of IPods and 20% of sales price of Television contribute to the profits of Jamshedpur Electronics. How much profit did the company earn in the month of January from Bistupur and Kadma from the two products?

(a) Rs 513265
(b) Rs 3033000
(c) Rs 36453621
(d) Rs 5230213
(e) Rs 5350000

Q4. For Jamshedpur Electronics Beginning inventory was 720 for Televisions and 1800 form IPods and Ending inventory was 840 for Televisions and 1920 for IPods in the month of January.

How many units of Televisions and IPods did Jamshedpur Electronics order for the month of January?

(a) 1020, 2270
(b) 1020, 2370
(c) 2270, 1030
(d) 9120, 31620
(e) 1020, 2280

Rough Work

Answer Grid

Q1	Q2	Q3	Q4
b	d	b	d

Solutions to Q1–Q4

A1. Unit sold = 3150 units
Inventory at beginning = 2520
Inventory at end = 2880
Number of order of TVs placed = (3150 – 2520) + 2880 = 3510
Value of order placed = 3510 × Rs 1150 = Rs 4036500
Hence, option (b) is the answer

A2. No. of Television sold in January = 9000
Sales value = 9000 × Rs 1150 = Rs 10350000
Surcharge = 14% of 10350000 = Rs 1449000
Hence, option (d) is the answer

A3. No. of Televisions sold = 7200
No. of IPod sold = 22050

Sales value of Television = 7200 × Rs 1150 = Rs 8280000
Sales value of IPod = 22050 × Rs 1000 = Rs 22050000
10% of sales value of Television = Rs 828000
20% of sales value of IPod = Rs 2205000
Therefore Total profit = Rs 2205000 + Rs 828000 = Rs 3033000
Hence, option (b) is the answer

A4. In January,
No. of Television sold = 9000 units
No. of IPod sold = 31500 units
Hence no. of Television whose order is placed in the month of January = (9000 + 840 – 720) = 9120
And no. of IPod's order = (31500 + 1920 – 1800) = 31620
Hence, option (d) is the answer

Employment Scenario Set 44

Direction for questions 1–5

Refer to the four pie charts related to the sectoral composition of output and their respective share in employment in 1970 and 1994:

Sectoral Composition of India's Employment in 1970

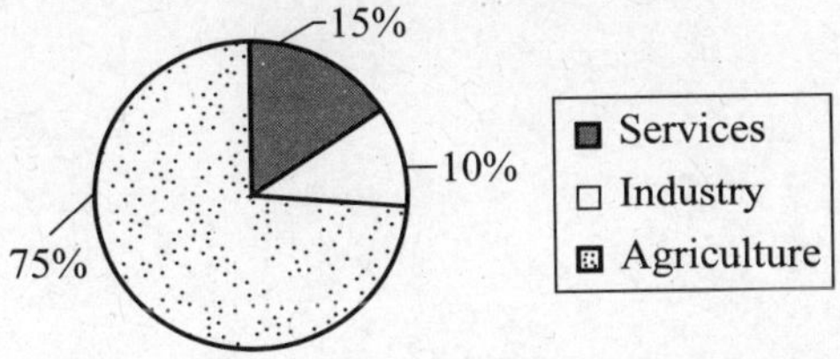

Sectoral Composition of India's Employment in 1994

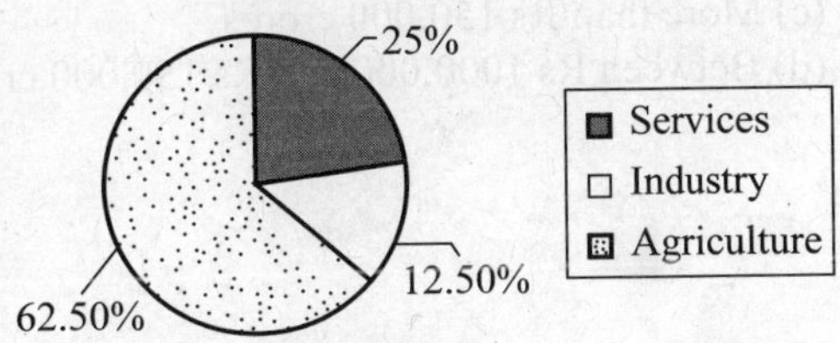

Sectoral Composition of India's Output in 1970 (% of GDP) GDP = Rs 42,222 crores

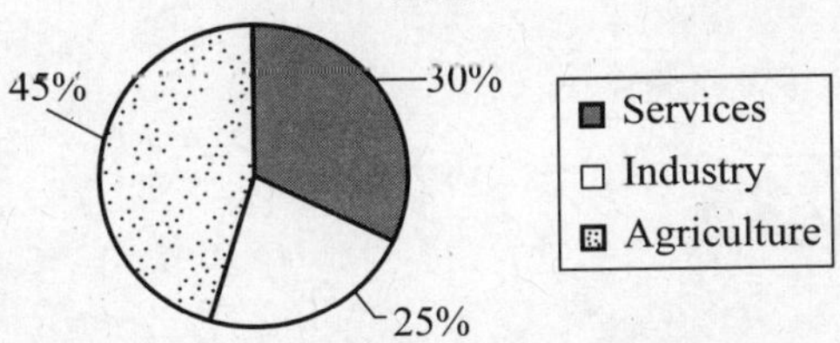

Sectoral Composition of India's Output in 1994 (% of GDP) GDP = Rs 570,954 crores

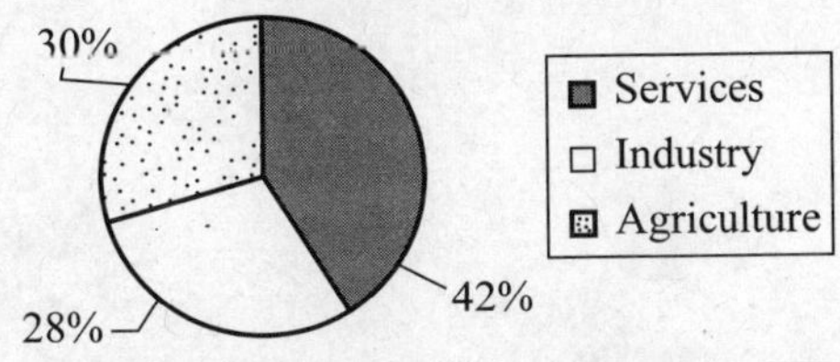

Q1. Which of the following statements is FALSE?

(a) While the share of agriculture in the GDP has come down, the share of services and industry has increased.

(b) Agriculture continues to be the predominant provider of employment in India in 1994.

(c) Percentage point increase in the share of employment in services is more than percentage point increase in its share in GDP.

(d) None of these

Q2. Between 1970 and 1994 the Indian GDP, in rupee terms

(a) Has increased only marginally in services

(b) Has increased maximum for services

(c) Has risen by more than 15 times for both services and industry

(d) None of these

Q3. Between 1970 and 1994 the GDP of India in agriculture

(a) Decreased by 15%

(b) Increased to approximately nine times

(c) Cannot be calculated based on the data given
(d) None of these

Q4. In 1970, the contribution to GDP from services in rupee terms was
(a) Less than Rs 10,000 crores
(b) More than Rs 15,000 crores
(c) Between Rs 10,000 and 15,000 crores
(d) None of these

Q5. In 1994 if the share of industry in output (GDP) was 30% instead of 28%, then in rupee terms, it would have been
(a) Less than Rs 100,000 crores
(b) More than Rs 200,000 crores
(c) More than Rs 150,000 crores
(d) Between Rs 1000,000 and Rs 150,000 crores

Rough Work

Answer Grid

Q1	Q2	Q3	Q4	Q5
c	b	b	c	c

Solutions to Q1–Q5

A1. Through observations, it can be seen that Options (a) and (b) are true.
Consider option (c) now:
Percentage point increase in the share of Services in Employment = 25% – 15% = 10%
Percentage point increase in the share of Services in GDP = 42% – 30% = 12%
So, Percentage point increase in the share of Services in Employment < Percentage point increase in the share of Services in GDP. Hence option (c) is false.
Hence option (c) is the answer

A2. Since overall value is increased i.e. 42,222 to 5,70,954

$$\% \text{ increase in Services} = \frac{42-30}{30}\times 100 = \frac{12}{30}\times 100 = 40\%$$

$$\% \text{ increase in Industry} = \frac{28-25}{25}\times 100 = 12\%$$

% of Agriculture is decreased from 45% to 30%
Therefore maximum increase is in services.
Hence option (b) is the answer.

A3. GDP of India in Agriculture in 1970 = 45% of 42,222 = 19,000 (approx)
GDP of India in Agriculture in 1994 = 30% of 5,70,954 = 1,71,000 (approx)
Hence, GDP of India in Agriculture has increased to 9 times approximately.
Hence option (b) is the answer.

A4. In 1970, contribution to GDP from Services = 30% of Rs 42,222 crores = Approximately Rs 12,600 crores
Hence option (c) is the answer.

A5. 30% of 5,70,954 = Approximately 1,71,000
Hence option (c) is the answer.

Arrival and Departure at T7

Set 45

Direction for questions 1–6

Go through the data set given below and solve the questions based on it.

Following table presents the data regarding the arrivals and departures at T7 airport.

Delay (minutes)	Average Delay (minutes)	No of Arrivals	No of Departures
0	–	120	155
0–30	15	40	30
30–60	45	30	10
> 60	–	10	5
Total		200	200

Q1. What % of the aircrafts that arrive and depart from the airport are late by an average of 45 minutes or more?

(a) 10% (b) 18.75% (c) 27% (d) 13.75%

Q2. What % of the total arrivals are late arrivals?

(a) 60% (b) 77.5% (c) 40% (d) 48%

Q3. What was the average delay (in minutes) of the arriving flights on that particular day?

(a) 9 min 45 seconds (b) 19 min 30 sec (c) 30 minutes (d) Cannot be determined

Q4. What was the average delay in minutes of the departing flights on that particular day?

(a) 9 minutes (b) 4 min 50 sec (c) 4 min 30 sec (d) Cannot be determined

Q5. What % of the flights that arrived and departed from the airport did so on time?

(a) 86.25% (b) 68.75% (c) 71.25% (d) Cannot be determined

Q6. What was the ratio of the number of late departures to that of late arrivals?

(a) 1: 1.77 (b) 1: 2.25 (c) 1.96: 1 (d) 2.89: 1

Rough Work

Answer Grid

Q1	Q2	Q3	Q4	Q5	Q6
d	c	d	d	b	a

Solutions to Q1–Q6

A1. Total arrivals = 200.
Number of arrivals late by an average of 45 minutes or more = 30 + 10 = 40.
Total departures = 200.
Number of departures late by an average of 45 minutes or more = 10 + 5 = 15
Total arrivals + Departures = 400.

Total late arrivals + late departures (45 min or more) = $\frac{55}{400} \times 100 = 13.75\%$.
Hence option (d) is the answer.

A2. Total arrivals = 200.
Total late arrivals = Total arrivals – number of arrivals delayed by 0 min = 200 – 120 = 80 of them arrived late.

Percentage of late arrivals = $\frac{80}{200} \times 100 = 40\%$.

Hence option (c) is the answer.

A3. We don't have the exact data for arrivals late by more than 60 minutes. We only know that these arrivals are more than 60 minutes. Hence average delay in arrivals cannot be calculated.
Hence option (d) is the answer.

A4. Explanation same as that of A3.
Hence option (d) is the answer.

A5. Total number of arrival and departures = 400.
Total arrival and departures on time = 120 + 155 = 275.

Therefore, % of arrival and departures on time = $\frac{275}{400} \times 100 = 68.75\%$.

Hence option (b) is the answer.

A6. Late departures: late arrivals = 45: 80 = 9: 16 = 1: 16/9 = 1: 1.77.
Hence option (a) is the answer.

Rang Barsey Paint Company

Set 46

Direction for question 1–3

Answer the questions on the basis of the information given below.

Rang Barsey Paint Company (RBPC) is in the business of manufacturing paints. RBPC buys red, yellow, white, orange, and pink paints. orange paint can also be produced by mixing red and yellow paints in equal proportions. Similarly, pink paint can also produced by mixing equal amounts of red and white paints. Among other paints, RBPC sells cream paint, (formed by mixing white and yellow in the ratio of 70:30) Avocado paint (formed by mixing equal amounts of orange and pink paint) and washed orange paint (formed by mixing equal amounts of orange and white paint).

The following table provides the price at which RBPC buys paints.

Color	Rs/Litre
RED	20
YELLOW	25
WHITE	15
ORANGE	22
PINK	18

Q1. The cheapest way to manufacture Avocado paint would cost

(a) Rs 19.50 per litre
(b) Rs 19.75 per litre
(c) Rs 20.00 per litre
(c) Rs 20.25 per litre

Q2. Washed Orange can be manufactured by mixing

(a) Cream and red in the ratio 14:10
(b) Cream and red in the ratio 3:1
(c) Yellow and pink in the ratio 1:1
(d) Red, yellow and white in the ratio 1:1:2

Q3. Assume that Avocado, cream and washed Orange each sells for the same price. Which of the three is the most profitable to manufacture?

(a) Avocado
(b) Cream
(c) Washed Orange
(d) Sufficient data is not available

Answer Grid

Q1	Q2	Q3
b	d	b

Solutions to Q1–Q3

A1. Paint Avocado is made by mixing Orange and Pink in equal quantities.
If Orange is made using Red and Yellow, then the cost of Orange would be $(20 + 25)/2 = 22.5$ which is greater than the cost of the ORANGE.

If we make Pink by Red and White, the cost of Pink would be $\frac{7\times15+3\times75}{10} = 17.5$ and it is less than the cost of the Pink paint.

Hence, the cost of the paint Avocado is $\frac{22+17.5}{2} = 19.75$

Hence option (b) is the answer.

A2. The ratio of red, yellow and white is 1:1:2.
Hence option (d) is the answer.

A3. Cost of Avocado paint is Rs 19.75.

Cost of Cream is $\frac{7\times15+3\times75}{10} =$ Rs 18

Cost of Washed Orange is Rs 18.50.
Since selling price is same, hence manufacturing cream is most profitable. Hence option (b) is the answer.
Hence option (b) is the answer.

Basics of Management Accounting

Set 47

Direction for questions 1–4

Go through the following information given below and solve the questions based on it.

Prof. Bansal has been teaching his students basic terminology involved in deciding the financial health of a company ABC Ltd. Go through the following charts given for ABC Ltd and answer the questions based on it.

Terms defined:

(a) Current Ratio = Current Assets/Current Liability

This ratio is mainly used to have an idea of the company's ability to pay back its short-term liabilities (debt and payables) with its short-term assets (cash, inventory, receivables).

Chart 1:

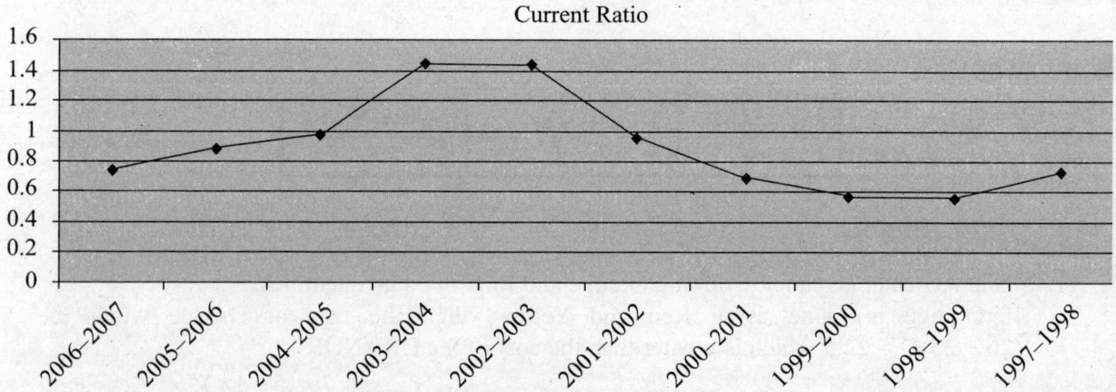

(b) Gross profit margin—Gross profit margin presents the percentage of revenue/sales value left after subtracting the cost of goods sold.

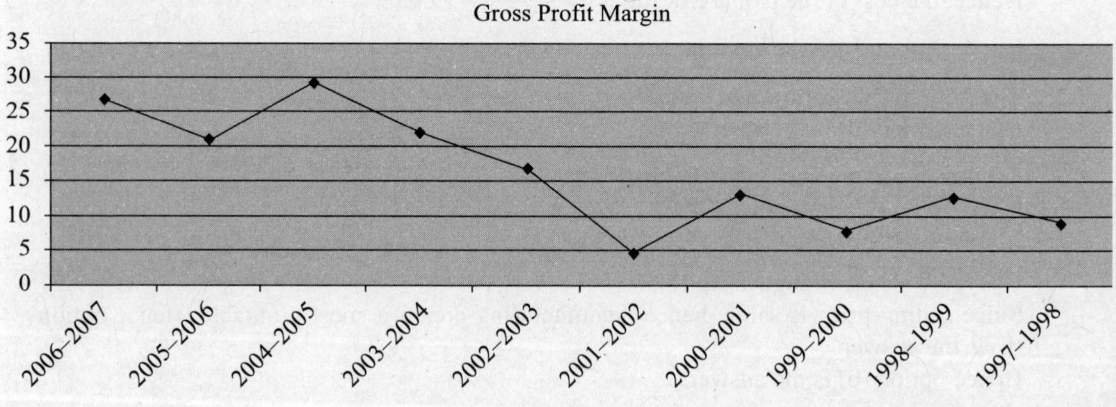

(C) Total Debt/Equity ratio—The debt-to-equity ratio offers one of the best pictures of a company's leverage.

A company without any debt may be missing an opportunity to increase earnings by financing projects that could have been used to generate a better return. A little debt can be good for a company's earnings. On the other hand, a high debt-to-equity ratio translates into higher risk for shareholders.

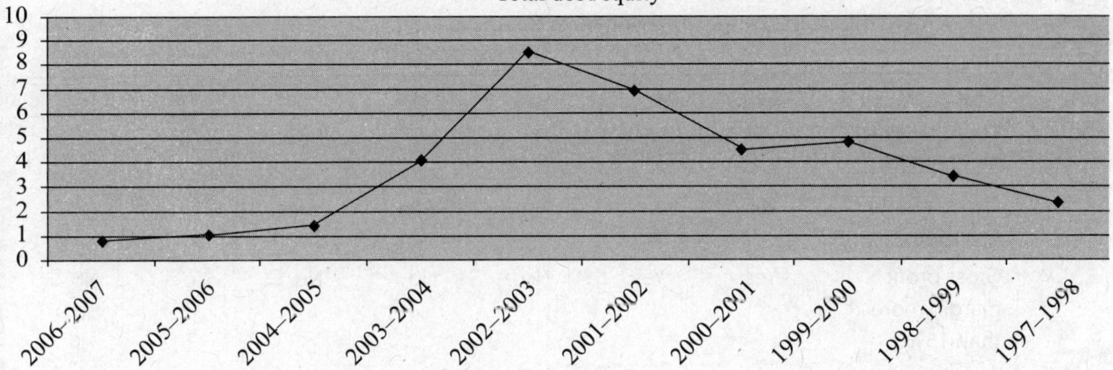

Q1. For how many years, company ABC Ltd. has a total debt to equity ratio more than 2 with current ratio less than 0.8 and gross profit margin more than 15%?

(a) 0 (b) 1 (c) 2 (d) 3

Q2. Assume that the total equity for the company ABC Ltd. is Rs 9 million in 2002–03 which is 1/4th of the total current asset for the same year. If current liability is to be serviced through total debt, how much surplus would be generated in 2002–03 after servicing current liability through total debt?

(a) Between Rs 35 million to Rs 40 million (b) Between Rs 40 million to Rs 45 million
(c) Between Rs 45 million to Rs 50 million (d) Between Rs 30 million to Rs 35 million

Q3. A Company is said to be "Going Northward" any year if its Gross profit margin as well as current ratio show the same trend over the previous year. For how many years company ABC Ltd. can be said to be "Going Northward"?

(a) 2 (b) 3 (c) 4 (d) None of these

Q4 If Gross profit margin is calculated over total current assets (instead of the current system of calculating over the total sales revenue minus cost of goods sold), which year will show the maximum Gross profit margin?

(a) 2005–06 (b) 2001–02 (c) 2004–05 (d) Cannot be determined

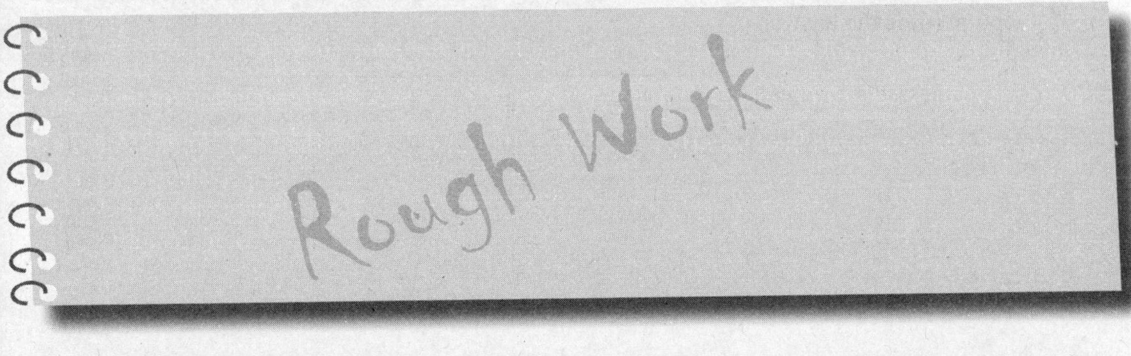

Answer Grid

Q1	Q2	Q3	Q4
a	b	c	d

Solutions to Q1–Q4

A1. Following table presents the situation clearly:

Year	03–04	02–03	01–02	00–01	99–00	98–99	97–98
Total Debt to Equity Ratio	4.15	8.59	7	4.57	4.87	3.41	2.35
Current Ratio less than 0.8	No	No	No	Yes	Yes	Yes	Yes
Gross profit margin more than 15%	Yes	Yes	No	No	No	No	No
Final Verdict	No	No	No	No	No	No	No

It can be inferred that none of the years satisfy all the three conditions. Hence option (a) is the answer.

A2. In 2002–03, total debt to equity ratio = 8.6 ⇒ Total debt = 8.6 × Rs 9 million = Rs 77.4 million.
Total current liabilities = Rs 9 million × 4 = Rs 36 million
After servicing the current liability through total debt, there would be a surplus (since Total debt > Total current liabilities). Surplus = Rs 77. 4 million – Rs 36 million = Rs 41.4 million. Hence option (B) is the answer.

A3. Following are the years:
2000–2001, 2002–03, 2003–04, 2005–06. So there are four years when Gross profit margin as well as current ratio have either both decreased or both increased over the previous year. Kindly note that data presents years in a right to left format.
Hence option (c) is the answer.

A4. There is no relation given between sales value minus cost of goods sold and total current asset. So Gross profit margin calculated over total current asset cannot be calculated. Hence option (d) is the answer.

Car Sales Data

Set 48

Direction for questions 1–4

Go through the data set given below and solve the questions based on it.
Following table presents the sales volume of the following car manufacturing companies in the month of August for the given two years:

Company	2011	2010
Maruti	77086	92874
Hyundai	26677	28601
Tata	16829	25312
Mahindra	15664	13796
Toyota	11693	6361
Ford	7392	7925
Honda	6987	3531
Skoda	1812	1512
Mercedes	614	573

Q1. Which car manufacturing company listed above has shown maximum percentage increase in the sales volume in 2011 August over 2010 August?
(a) Toyota (b) Skoda (c) Honda (d) None of these

Q2. How many companies have shown a growth in the sales volume in 2011 as compared to 2010?
(a) 4 (b) 5 (c) 6 (d) Cannot be determined

Q3. Which of the following statements is true for the month of August?
(a) Number of companies seen the growth in sales volume > Number of companies seen a decrease in sales volume
(b) Number of companies seen the growth in sales volume < Number of companies seen a decrease in sales volume
(c) Number of companies seen the growth in sales volume = Number of companies seen a decrease in sales volume
(d) Cannot be determined

Q4. For the month of August 2011, sales volume of which company is closest to the average sales value of August 2011?
(a) Hyundai (b) Tata (c) Mahindra (d) Cannot be determined

Rough Work

Answer Grid

Q1	Q2	Q3	Q4
c	d	a	d

Solutions to Q1–Q4

A1. In this question, with a bit of guestimation, answer can be calculated easily.
Following companies have shown growth: Mahindra, Toyota, Honda, Skoda and Mercedes.
Out of these companies, Toyota and Ford have become almost twice, and clearly one of these two is going to be the answer.
Observing the data for Toyota and Ford, we can observe that in case of Honda, August 2011 value is almost equal to twice the value of August 2010, whereas in case of Toyota, it is slightly less than the twice the value of August 2010.
Hence option (c) is the answer.

A2. Data given in the table is for August month only, whereas question is asking for the whole year. Hence question cannot be answered.
Hence option (d) is the answer.

A3. Number of companies seen the growth in sales volume = 5
Number of companies seen a decrease in sales volume = 4
Hence option (a) is the answer.

A4. Table given in the question set gives the data for only these companies. There may be more companies manufacturing cars. It is also not given these companies are the top companies.
Hence option (d) cannot be determined is the answer.

Air ACs

Set 49

Direction for questions 1–4

Go through the charts given below and solve the questions based on it:

Line chart given below presents the number of ACs sold by Air ACs from the year 2009 to 2013. Data for 2014 is projection. All the data points are in the multiple of 500.

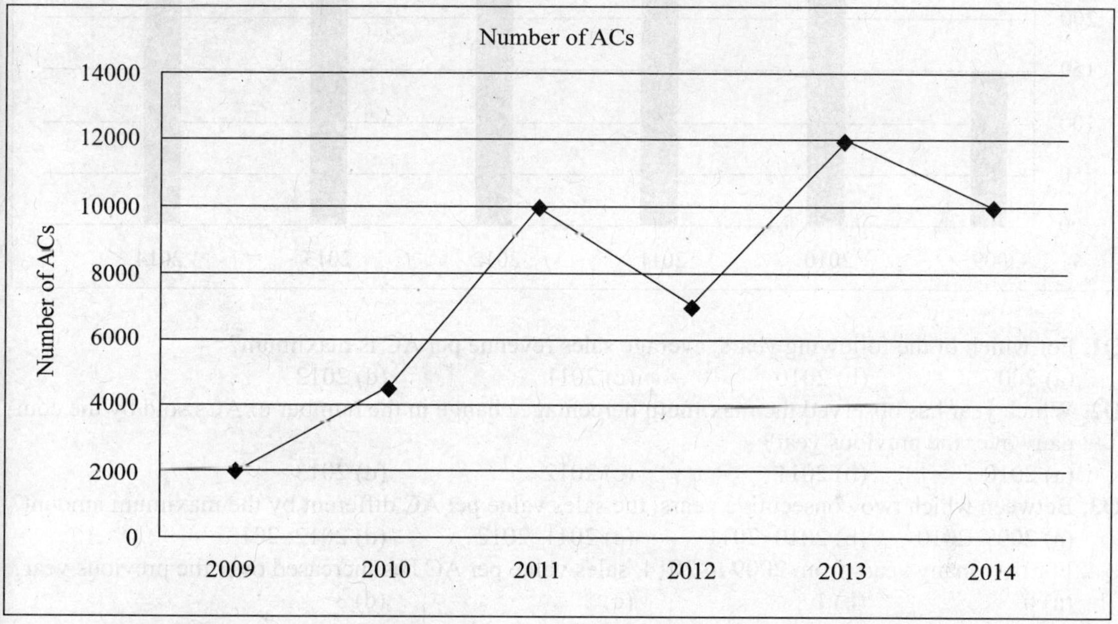

Bar chart given below presents the sales value obtained by Air ACs by selling the number of units as mentioned in the line chart from the year 2009 to 2013. Data for 2014 is projection on the basis of previous year's data.

Assume that everything produced is sold in the same year. Also assume that there is no import or export.

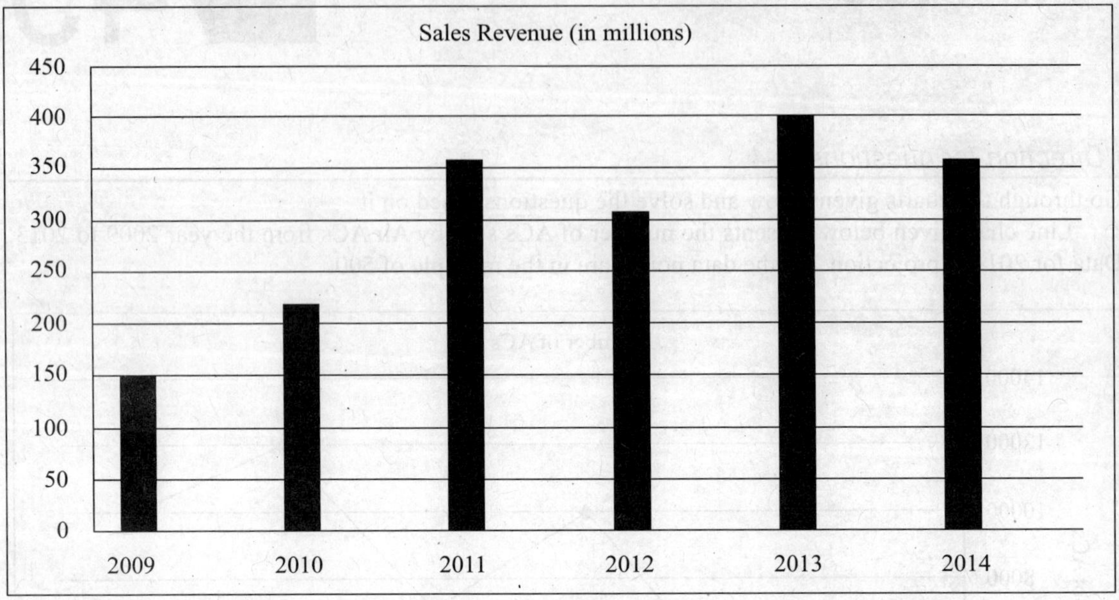

Q1. For which of the following years, average sales revenue per AC is maximum?
(a) 2009 (b) 2010 (c) 2011 (d) 2012

Q2. Which year has observed the maximum percentage change in the number of ACs sold by the company over the previous year?
(a) 2010 (b) 2011 (c) 2012 (d) 2013

Q3. Between which two consecutive years, the sales value per AC different by the maximum amount?
(a) 2009–2010 (b) 2010–2011 (c) 2011–2012 (d) 2012–2013

Q4. For how many years from 2009 to 2014, sales value per AC has increased over the previous year?
(a) 0 (b) 1 (c) 2 (d) 3

Rough Work

Answer Grid

Q1	Q2	Q3	Q4
a	a	a	c

Solutions to Q1–Q4

A1. Options mention four particular years, so we should be evaluating only these four years.

Year	Average Sales Revenue (in millions)
2011	0.036
2012	0.044
2010	0.048
2009	0.075

Hence option (a) is the answer.

A2. Question is asking for "Percentage change" – it can be percentage increase or percentage decrease. Since it can be seen that there are data points that see percentage increase of more than 100%, we are not required to see for percentage decrease as there cannot be a percentage decrease of more than 100%.

So the likely answers are—option (a), option (b) and option (c). Option (c) is too low in comparison with options (a) and (b). In option (a) and (b), the data for the year is double the data for the previous year, which is not in case of option (c). Hence we rule out option (c).

Evaluating options (a) and (b):

Option (a) – Percentage increase $= \dfrac{4500-2000}{2000} \times 100 = \dfrac{5500}{4500} \times 100 = 122\%$

How to do it fast: Convert the numbers into simple ratios:

$$\frac{2500}{2000} \times 100 = \frac{5}{4} \times 100 = 125\%$$

Option (b) - Percentage increase $= \dfrac{1000-4500}{4500} \times 100 = \dfrac{5500}{4500} \times 100 = 122\%$

How to do it fast: Convert the numbers into simple ratios:

$$\frac{11}{9} \times 100 = \frac{9+2}{9} \times 100 = 122\%$$

$$[\frac{9+2}{9} = \frac{9}{9} + \frac{2}{9}] \times 100 = 100 + 22.22 = 122.22\%$$

Hence option (a) is the answer.

A4. It happens only in 2012 (over 2011) and in 2014 (over 2013). Hence option (c) is the answer.

The Score So Far

Set 50

Direction for questions 1–5

Go through the table given below and solve the questions based on it.

Following table provides the data regarding the cricket matches played. All the matches are played between two teams. Table also mentions name of teams, matches played by them, and number of matches won/lost/tied and No result.

Team	Matches Played	Matches Won	Matches Lost	Matches Tied	No Result
Australia	811	500	275	9	27
Pakistan	780	419	338	6	17
India	817	405	371	6	35
West Indies	687	353	304	6	24
Sri Lanka	681	318	330	4	29
South Africa	478	297	163	5	13
England	592	288	277	7	20
New Zealand	632	271	323	5	33
Zimbabwe	407	107	286	5	9
Bangladesh	267	75	190	0	2
Kenya	146	39	102	0	5
Ireland	74	34	35	1	4
Netherlands	69	26	41	0	2
Scotland	55	19	33	0	3
Canada	71	17	53	0	1
Afghanistan	23	12	11	0	0
Bermuda	35	7	28	0	0
Asia XI	7	4	2	0	1
ICC world XI	4	1	3	0	0
Africa XI	6	1	4	0	1
UAE	11	1	10	0	0
USA	2	0	2	0	0
East Africa	3	0	3	0	0
Hong Kong	4	0	4	0	0
Namibia	6	0	6	0	0

Q1. Winning percentage =

$$\frac{\text{Number of matches won}}{(\text{Total number of matches played} - \text{number of matches tied} - \text{no result matches})} \times 100$$

For how many teams, the winning percentage is more than 50%?

(a) 3 (b) 6 (c) 8 (d) None of these

Q2. Winning percentage = $\frac{\text{Number of matches won}}{(\text{Total number of matches played})} \times 100$

Which team has the largest winning percentage?

(a) Australia (b) South Africa (c) Pakistan (d) Asia XI

Q3. Considering that these are the only teams that play cricket, what is the total number of matches played?

(a) 6668 (b) 6688 (c) 3334 (d) 3344

Q4. Considering that these are the only teams that play cricket, what percentage of the matches have ended up with "no result"?

(a) Between 2%–3% (b) Between 3%–4% (c) Between 4%–5% (d) None of these

Q5. What is the total number of matches tied?

(a) 54 (b) 44 (c) 27 (d) 22

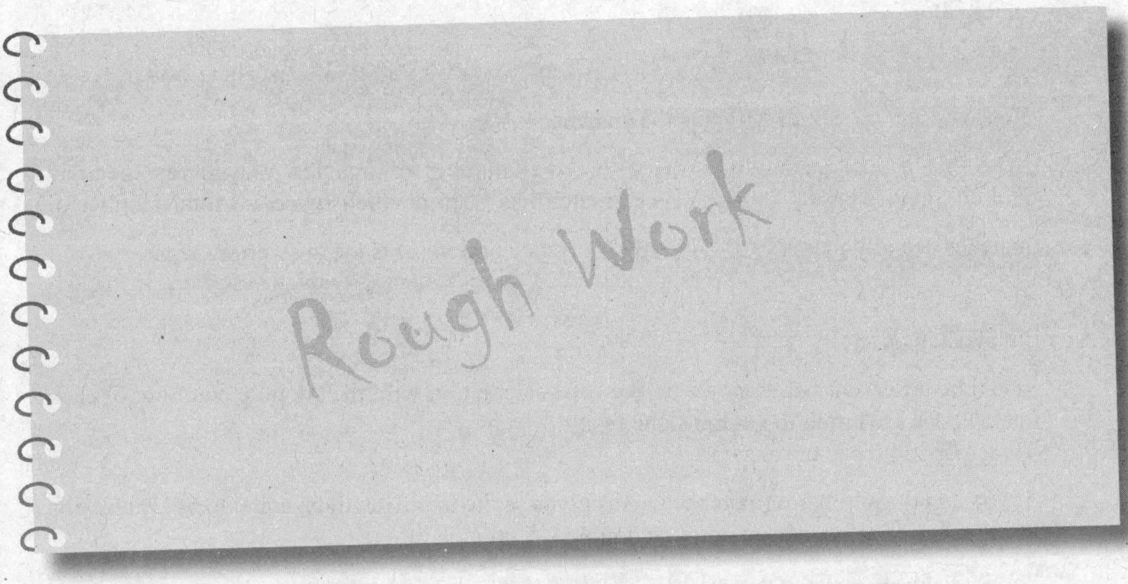

Answer Grid

Q1	Q2	Q3	Q4	Q5
c	b	c	b	c

Solutions to Q1–Q5

A1. It can be clearly seen that to have a winning percentage of more than 50%, number of matches won shall be greater than half the [number of matches played (as given in the 1st column) – number of matches tied – No result matches].
Winning percentage is more than 50% for the following countries - Australia, Pakistan, India, West Indies, South Africa, England, Afghanistan, Asia XI. Hence option (c) is the answer.

A2. Answering this question would have been slightly more difficult had the fourth option been "None of These" instead of Asia XI. Since options mention four teams namely, we will check for these four teams only.
Note from Author – Had the fourth option been "None of these", it would have involved a lot of calculation. In that case, my suggestion is - Attempt this question at the end – once you have seen all the questions.
Winning percentage for the mentioned teams is as follows:

Australia	61.60%
Pakistan	53.70%
South Africa	62.10%
Asia XI	57.10%

Hence option (b) South Africa is the answer.

A3. Now this is just an addition exercise. Total number of matches played (as given in 2nd column) = 6668. Though, every match has been counted twice, so total number of matches actually played $= \frac{6668}{2} = 3334$. Hence option (c) is the answer.

A4. Final calculation $= \frac{226}{6668} \times 100 = 3.33\%$

It can be observed that even we do the final calculation with the "double counting of each match", we are going to get the same result.
Hence option (b) is the answer.

A5. If you add the number of matches tied as given in the table, it will be equal to 54. But in this case, every match has been counted double.

So, total number of matches tied $= \frac{54}{2}$
Hence option (c) is the answer.

Asylum Seekers

Set 51

Direction for questions 1–5

Read the information given below and solve the questions based on it.

Following table provides the data regarding the people who are seeking asylum in different countries.

Asylum seekers: Share of top 15 receiving countries in total number of applications

All the data given in the table below are in percentages. For example, in 2007, USA received 15% application out of total asylum seekers applications.

Countries	2007	2008	2009	2010	2011
USA	15	13	13	15	17
France	9	9	11	13	12
Germany	6	6	7	11	10
Italy	4	8	5	3	8
Sweden	11	6	6	9	7
Belgium	3	3	5	6	6
UK	8	8	8	6	6
Canada	8	10	9	6	6
Switzerland	3	4	4	4	4
Turkey	2	3	2	3	4
Austria	4	3	4	3	3
Netherlands	2	4	4	4	3
Australia	1	1	2	3	3
Greece	8	5	4	3	2
Norway	2	4	5	3	2

Q1. USA received 72800 applications in the year 2010. What is the total number of asylum seekers who didn't apply to any to the countries mentioned in the table (approximately)?

(a) 37800 (b) 36400 (c) 38800 (d) 40200

Q2. In how many years, sum of percentage of applications received by the top three rank countries is more than the sum of percentage of applications received by the remaining countries in the table?

(a) 1 (b) 2 (c) 3 (d) None of these

Q3. What is the sum of ranks of Germany across the given duration?

(a) 22 (b) 23 (c) 24 (d) 25

Q4. Between 2007 and 2011, the United States of America received the largest number of asylum-seekers = 278800 applications. What is the total number of asylum seekers' application received by France during the same period (approximately)?

(a) 216000 (b) 262800 (c) 243900 (d) Cannot be determined

Q5. In the table given above, which country has seen the minimum variation in 2011 over 2007 in the ranking as the destination for the asylum seekers?
(a) USA (b) France (c) Germany (d) UK

Rough Work

Answer Grid

Q1	Q2	Q3	Q4	Q5
c	d	c	d	a

Solutions to Q1–Q5

A1. In 2010, 92% of the total asylum seekers put forward their application to one of these 15 countries mentioned in the table. (See the solution A2).
So, 15% = 72800 ⇒ 8% (100% – 92%) = 38826. Hence option (c) is the answer.

A2. Since this table does not provide 100% data, and data for only top 15 countries, we are required to find out the summation of all the countries. Later on, we will see if half of this is contributed by the top three countries?
This is the total percentage applications received by these 15 countries:

Year	2007	2008	2009	2010	2011
Percentage obtained by top 15 countries	86	87	89	92	93

Percentage applications by top three countries

In 2007:
15% + 11% + 9% = 35% < half of 86%. Hence not the answer.
In 2008:
13% + 10% + 9% = 32% < half of 87%. Hence not the answer.
In 2009:
13% + 11% + 9% = 33% < half of 89%. Hence not the answer.
In 2010:
15% + 13% + 11% = 39% < half of 92%. Hence not the answer.
In 2011:
17% + 12% + 10% = 39% < half of 93%. Hence not the answer.
So, in none of the years, sum of percentage of applications received by the top three rank countries is more than the sum of percentage of applications received by the remaining countries in the table.
Hence option (d) is the answer.

A3. Following is the ranking of the Germany:

Countries	2007	2008	2009	2010	2011
Germany	7	6	5	3	3

Hence sum of the ranks = 24. Hence option (c) is the answer.

A4. We don't know the year wise breakup of the data for the applications received by USA. Hence year wise data for France cannot be calculated. Hence option (d) cannot be determined is the answer.

A5. USA is consistently ranked 1 as the favorite destination of the asylum seekers (as it receives the maximum percentage of applications in all the years). Hence option (a) is the answer.

Sales vs Net Profit

Set 52

Direction for questions 1–4

Go through the data set given below and solve the questions based on it.

The figure below represents states sales and net profit in Rs crore of a company for five years from 1994–95 to 1998–99. During this period the sales increased from Rs 100 crore to Rs 680 crore. Correspondingly, the net profit increased from Rs 2.5 crore to Rs 12 crore.

Net profit is defined as the excess of sales over total costs.

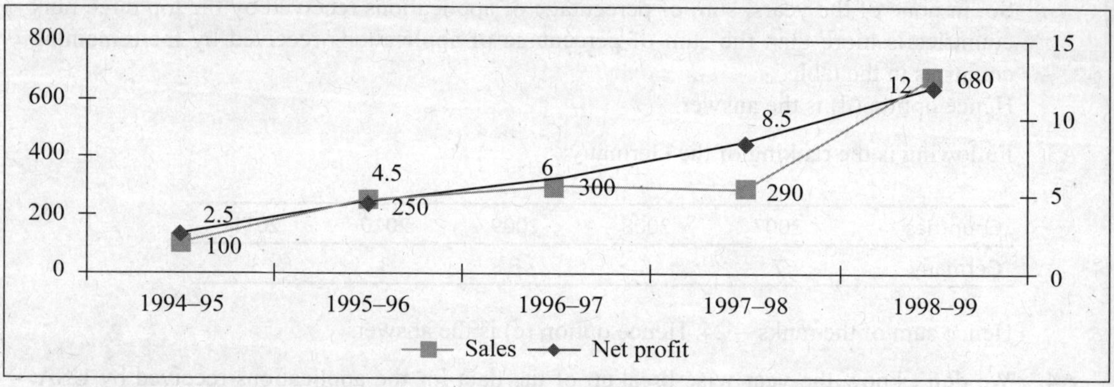

Q1. The highest percentage of growth in sales, relative to the previous year occurred in

(a) 1995–96 (b) 1996–97 (c) 1997–98 (d) 1998–99

Q2. The highest percentage growth in net profit, relative to the previous year, was achieved in

(a) 1998–99 (b) 1997–98 (c) 1996–97 (d) 1995–96

Q3. Defining the profitability as the ratio of net profit to sales, IVP Ltd, recorded the highest profitability in

(a) 1998–99 (b) 1997–98 (c) 1994–95 (d) 1996–97

Q4. With profitability as defined in question 3, it can be concluded that:

(a) Profitability is non-decreasing during the five years from 1994–95 to 1998–99.

(b) Profitability is non-decreasing during the five years from 1994–95 to 1998–99.

(c) Profitability remained constant during the five years from 1994–95 to 1998–99

(d) None of above.

Rough Work

Answer Grid

Q1	Q2	Q3	Q4
a	d	b	d

STORYLINE

Question set provides two line charts—one providing the sales value and other one presenting the data for profit for five years. You may expect questions related to profitability or average profit or sales. These questions will required to do simple calculations like additions, averages, percentage calculations and ratio comparison.

Solutions to Q1–Q3

A1. Looking at the line chart, it can be seen that answer has to be one of the two years – either 1995–96 or 1998–99. Instead of calculating percentage increase in sales, we can simply see the multiplier ratio.

For 1995–96: Multiplier $= \frac{Sales\ in 1995-96}{Sales\ in 1994-95} = \frac{250}{100} = 2.5$

For 1998–99: Multiplier $= \frac{Sales\ in 1998-99}{Sales\ in 1997-98} = \frac{680}{290} = \frac{580+100}{290}$ = 2 + (less than 0.5) = Less than 2.5

Hence 1995–96 shows the maximum percentage increase in sales over previous year. Hence option (a) is the answer.

A2. You can use the same method as used in Q1.

In case of net profit, the largest multiplier is obtained in case of 1995–96.

For 1995–96: Multiplier $= \frac{Profit\ in 1995-96}{Profit\ in 1994-95} = \frac{4.5}{2.5} = 1.8$

For all the other years, multiplier is less than this. Hence option (d) is the answer.

A3. This question requires simple calculation. For 1997–98, the ratio is maximum. Hence option (b) is the answer.

Marks Obtained in an Examination

Set 53

Direction for questions 1–5

Go through the data set given below and solve the questions based on it.

Following table gives the marks obtained by 6 students A, B, C, D, E and F in 6 subjects—History, Science, Geography, English, Hindi and Math. Marks in bracket show total marks in that subject. For example—History (150) shows that total marks in History = 150.

Marks obtained by 6 students in 6 subjects							
Students/ Subjects	History (150)	Science (200)	Geography (150)	English (200)	Hindi (100)	Math (200)	Total (1000)
A	75	110	90	140	75	170	660
B	105	130	75	130	80	140	660
C	95	105	80	150	90	160	680
D	85	115	95	125	65	135	620
E	115	135	110	145	70	125	700
F	120	160	96	110	55	145	686

Q1. How many students have got more than 50% marks in each of the subjects and more than 65% marks in total?

(a) 0 (b) 1 (c) 2 (d) 3 (e) None of these

Q2. What is the difference between the percentage marks obtained by the student B in History and percentage marks obtained by C in Hindi?

(a) 25 (b) 20 (c) 35 (d) 30 (e) None of these

Q3. For C, what is the difference between the percentage of marks in English and the percentage of marks obtained in total?

(a) 82 (b) $38\frac{1}{3}$ (c) 7 (d) 14 (e) None of these

Q4. Distinction is awarded if marks obtained is more than or equal to 75% in any subject. What is the sum total of number of distinctions awarded to all the students in all the subjects?

(a) 7 (b) 8 (c) 9 (d) 10 (e) None of these

Q5. In how many subjects did D get more than 70% marks?

(a) 0 (b) 1 (c) 2 (d) 3 (e) None of these

Rough Work

Answer Grid

Q1	Q2	Q3	Q4	Q5
d	b	c	c	a

STORYLINE

Question provides the data regarding the marks obtained by six students in six subjects. Total marks of all the subjects are not same.

Solutions to Q1–Q5

A1. C, E and F are the students. [We have to consider more than 50%]. Hence option (d) is the answer.

A2. $B \rightarrow \frac{105}{150} \times 100 = 70\%$

$C \rightarrow = 90\%$

Hence required difference = 90% – 70% = 20%. Hence option (b) is the answer.

A3. Percentage of marks of English

$= \frac{150}{200} \times 100\% = 75\%$

Average percentage of marks $= \frac{680}{1000} \times 100\% = 68\%$

Therefore required = 75 – 68 = 7. Hence option (c) is the answer.

A4. Look at the table given below:

	Total marks	Distinction marks	Total distinctions
History	150	112.5	2
Science	200	150	1
Geography	150	112.5	0
English	200	150	1
Hindi	100	75	3
Math	200	150	2
		Total	9

It can be seen that total distinctions = 9. Hence option (c) is the answer.

A5. In none of the papers, D get more than 70%. Hence option (a) is the answer.

Population in a Society Set 54

Direction for questions 1–5

Go through the data set given below and solve the questions based on it.

Following tables provide the population data of a society under different heads:

Number of Males and Females staying in various Societies		
Societies	Males	Females
A	250	350
B	400	150
C	300	275
D	280	300
E	180	250
F	325	300

Percentage of Male and Female Children in the Societies:

Societies	Children (out of total population)	Male children (out of total children)	Female children (out of total children)
A	25%	40%	60%
B	40%	75%	25%
C	16%	25%	75%
D	25%	80%	20%
E	40%	50%	50%
F	24%	46%	54%

Q1. What is the respective ratio of the number of adult females to the total number of female children staying in all the societies together?
(a) 243:82 (b) 112:71 (c) 82:243 (d) 71:112 (e) None of these

Q2. What is the total number of female children staying in all the societies together?
(a) 314 (b) 433 (c) 410 (d) 343 (e) None of these

Q3. What is the respective ratio of the total number of adult males in societies A and B together to the total number of adult males in societies E and F together?
(a) 3520 (b) 3360 (c) 4100 (d) 3000 (e) None of these

Q4. What is the total number of members staying in all the societies together?
(a) 3520 (b) 3360 (c) 4100 (d) 3000 (e) None of these

Q5. What is the difference between the number of male children in society B and the number of male children in society F?
(a) 84 (b) 14 (c) 96 (d) 26 (e) None of these

Rough Work

Answer Grid

Q1	Q2	Q3	Q4	Q5
a	c	d	b	c

STORYLINE

Question provides the data on three layers—Total population (number of males + number of females) ⇒ then the population of children as a percentage of total population ⇒ and finally out of the total children population, the number of male children and female children.

This set requires you to calculate a lot. You should expect such kind of set in online CAT.

Solutions to Q1–Q5

A1. Number of children:

Society A → $600 \times \dfrac{25}{100} = 15$

Number of female children = $150 \times \dfrac{60}{100} = 90$

Number of male children = $150 - 90 = 60$

Society B → $550 \times \dfrac{40}{100} = 220$

Number of female children = $220 \times \dfrac{25}{100} = 55$

Number of male children = $220 - 55 = 165$

Society C → $575 \times \dfrac{16}{100} = 92$

Number of female children = $92 \times \frac{75}{100} = 69$

Number of male children = 92 – 69 = 23

Society D → $580 \times \frac{25}{100} = 145$

Number of female children = $145 \times \frac{20}{100} = 29$

Number of male children = 145 – 29 = 116

Society E → $340 \times \frac{40}{100} = 172$

Number of female children = $172 \times \frac{50}{100} = 86$

Number of male children = 172 – 86 = 86

Society F → $625 \times \frac{24}{100} = 150$

Number of female children = $150 \times \frac{54}{100} = 81$

Number of male children = 150 – 81 = 69

Number of adult females = (350 + 150 + 275 + 300 + 250 + 300) – 410 = 1625 – 410 = 1215

Number of female children = 90 + 55 + 69 + 29 + 86 + 81 = 410

Hence required ratio = 1215:410 = 243:82

Hence option (a) is the answer.

A2. Using the data obtained in the previous question:
Number of all female children = 410. Hence option (c) is the answer.

A3. Number of adult males in societies A and B = (250 – 60) + (400 – 165) = 190 + 235 = 425
Number of adult males in societies E and F = (180 – 86) + (325 – 69) = 94 + 256 = 350
Required ratio = 425 : 350 = 17:14
Hence option (d) is the answer.

A4. Number of all members = (250 + 350) + (400 + 150) + (300 + 275) + (280 + 300) + (180 + 250) + (325 + 300) = 3360
Hence option (b) is the answer.

A5. Required difference = 165 – 69 = 96
Hence option (c) is the answer.

Gender Ratio

Set 55

Direction for questions 1–5

Go through the data set given below and solve the questions based on it.

Following pie chart gives the percentage of employees working in different departments of an organization. Total number of employees is 2500.

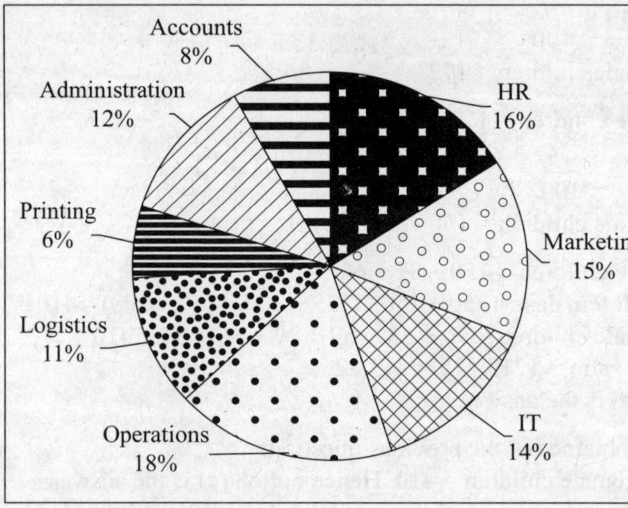

Following table gives the ratio of male and female employees in those departments in the same organization:

Department	Male	:	Female
Administration	7	:	5
Accounts	2	:	3
HR	5	:	3
Marketing	7	:	8
IT	3	:	4
Operations	5	:	4
Logistics	6	:	5
Printing	2	:	1

Assume all the data and questions are related to the same year.

Q1. What is the ratio between male employees in Administration and Printing Departments respectively?

(a) 7:4 (b) 4:7 (c) 3:4 (d) 7:3 (e) None of these

Q2. What is the difference between total number of employees in IT and Operations Departments?

(a) 75 (b) 150 (c) 100 (d) 50 (e) None of these

Q3. What is the ratio between total number of males in HR and Marketing and the total number of females in these two departments respectively?
(a) 13:15 (b) 15:13 (c) 13:17 (d) 17:14 (e) None of these

Q4. How many female employees are there in HR Department?
(a) 250 (b) 120 (c) 125 (d) 150 (e) None of these

Q5. What is the difference between the male and female employees in Logistics Department?
(a) 50 (b) 25 (c) 75 (d) 100 (e) None of these

Rough Work

Answer Grid

Q1	Q2	Q3	Q4	Q5
a	c	d	d	b

STORYLINE

Question set provides the data on two layers: Number of employees on the department level and then the ratio of males and females in every department. Questions will require a student to relate the data on these two levels.

Solutions to Q1–Q5

A1. Total number of employees in administration department $= \dfrac{2500 \times 12}{100} = 300$

Number of male employees $= \dfrac{7}{12} \times 300 = 175$

Total number of employees in printing department $= \dfrac{2500 \times 6}{100} = 150$

Number of male employees = $\frac{2}{3} \times 150 = 100$
Hence required ratio = 175:100 = 7:4
Hence option (a) is the answer.

A2. Required difference = $2500 \times (18 - 14)\% = \frac{2500 \times 4}{100} = 100$
Hence option (c) is the answer.

A3. Total number of employees in HR department $= \frac{2500 \times 16}{100} = 400$
Hence number of males = $\frac{5}{8} \times 400 = 250$
So number of females = 400 – 250 = 150
Number of employees in marketing department = $\frac{2500 \times 15}{100} = 375$
Number of males = $\frac{7}{15} \times 375 = 175$
Number of females = 375 – 175 = 200
Hence required ratio = (250 + 175) : (150 + 200) = 425:350 = 17:14
Hence option (d) is the answer.

A5. Total number of employees in logistics department = $\frac{2500 \times 11}{100} = 275$
Number of males = $\frac{6}{11} \times 275 = 150$
Number of females = 275 – 150 = 125
Hence required difference = 150 – 125 = 25
Hence option (b) is the answer.

Break-up of Population in Five Colonies

Set 56

Direction for questions 1–5

Go through the data set given below and solve the questions based on it.

Following bar chart presents the population data of five colonies—A, B, C, D and E. Population has been given in terms of percentage of adult males of total population, percentage of adult females of the total and percentage of children out of total population living in these colonies in a particular year. Assume all the questions pertain to this particular year only.

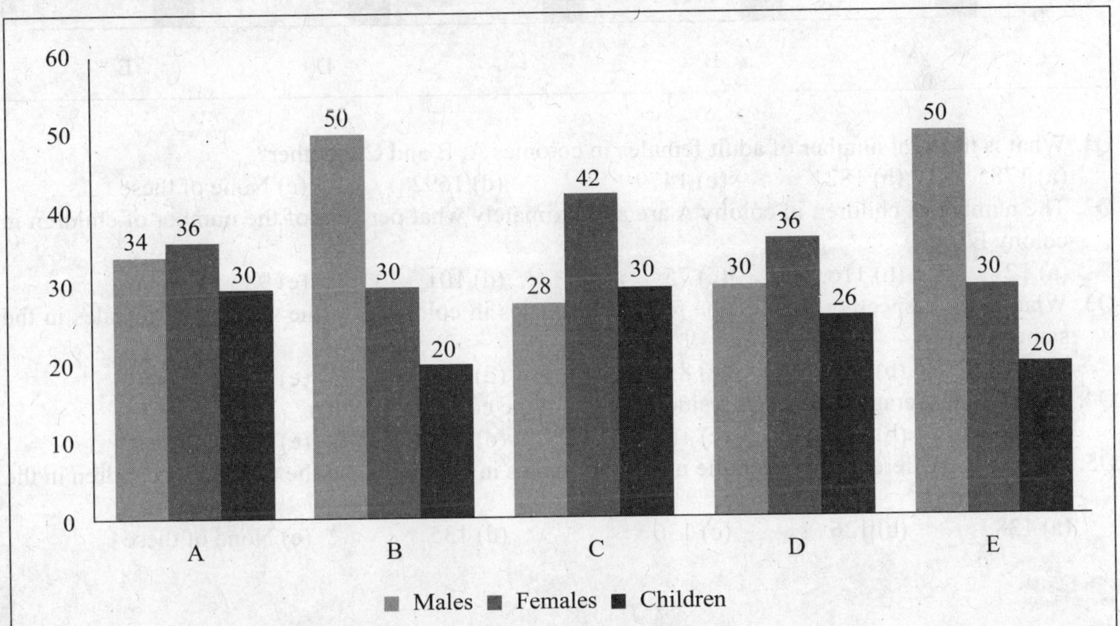

Following bar chart gives the total population of these five colonies in the same year.

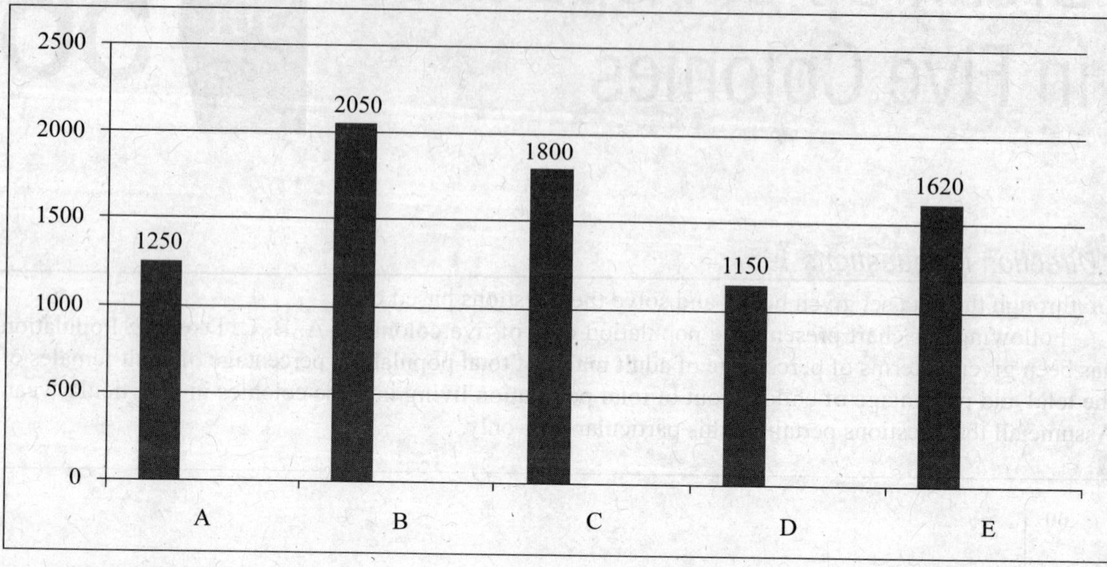

Q1. What is the total number of adult females in colonies A, B and C together?
(a) 1785 (b) 1821 (c) 1479 (d) 1692 (e) None of these

Q2. The number of children in colony A are approximately what per cent of the number of children in colony E?
(a) 121 (b) 116 (c) 75 (d) 101 (e) 98

Q3. What is the respective ratio of the number of males in colony B to the number of females in the same colony?
(a) 3:5 (b) 7:5 (c) 8:7 (d) 5:7 (e) None of these

Q4. What is the average number of residents from all the colonies together?
(a) 1654 (b) 1600 (c) 1580 (d) 1574 (e) None of these

Q5. What is the difference between the number of males in colony D and the number of children in the same colony?
(a) 138 (b) 126 (c) 136 (d) 135 (e) None of these

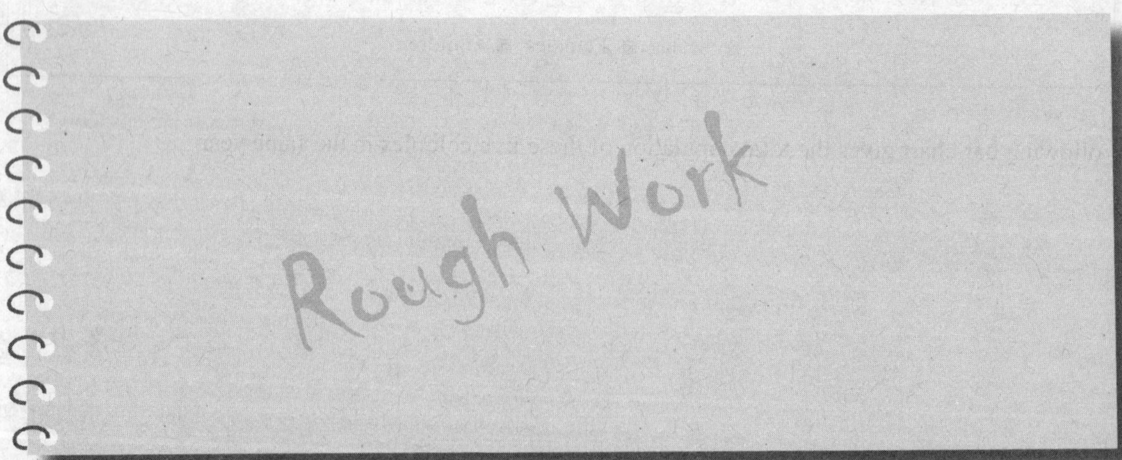

Answer Grid

Q1	Q2	Q3	Q4	Q5
b	b	e	d	a

STORYLINE

Question set provides the data on two layers: population data divided in three categories in terms of percentage of total population and total population. Questions will require a student to relate the data on these two levels.

Solutions to Q1–Q5

A1. Total number of adult females in colonies A, B and C together =

$$\left(\frac{1250 \times 36}{100} + \frac{2050 \times 30}{100} + \frac{1800 \times 42}{100}\right)$$

$= (450 + 615 + 756) = 1821.$

Hence option (b) is the answer.

A2. Number of children in colony A $= \frac{1250 \times 30}{100} = 375$

Number of children in colony E $= \frac{1620 \times 30}{100} = 324$

Required percentage $= \frac{375}{324} \times 100 = 116$

Hence option (b) is the answer.

A3. Required ratio = 50:30 = 5:3. Hence option (e) is the answer.

A4. Average no. of residents from all the colonies together $= \frac{1250 + 2050 + 1800 + 1150 + 1620}{5}$

$= \frac{7870}{5} = 1574$

Hence option (d) is the answer.

A5. Required difference = (38 – 26)% of 1150 $= \frac{12 \times 1150}{100} = 138$

Hence option (a) is the answer.

Two Pie Charts

Set 57

Direction for questions 1–5

Go through the data set given below and solve the questions based on it.

Following pie charts provide the percentage composition of students from a state enrolled in different faculty in 2001 and 2002.

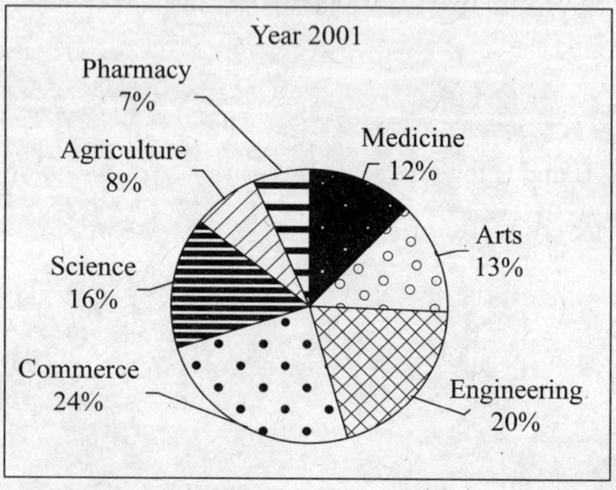

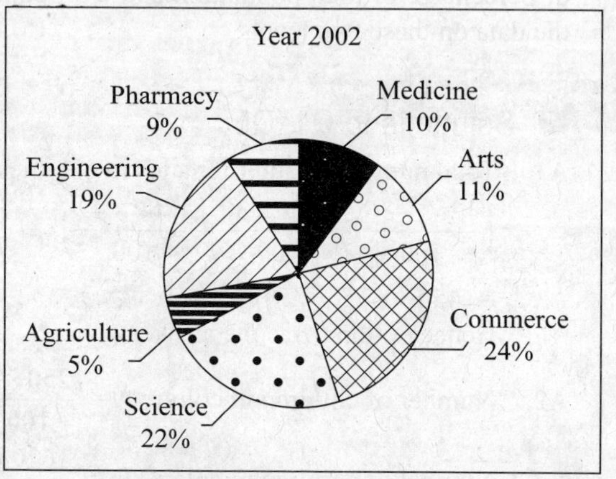

Following bar chart provides the total number of students from that particular state across four years – 2001 to 2004.

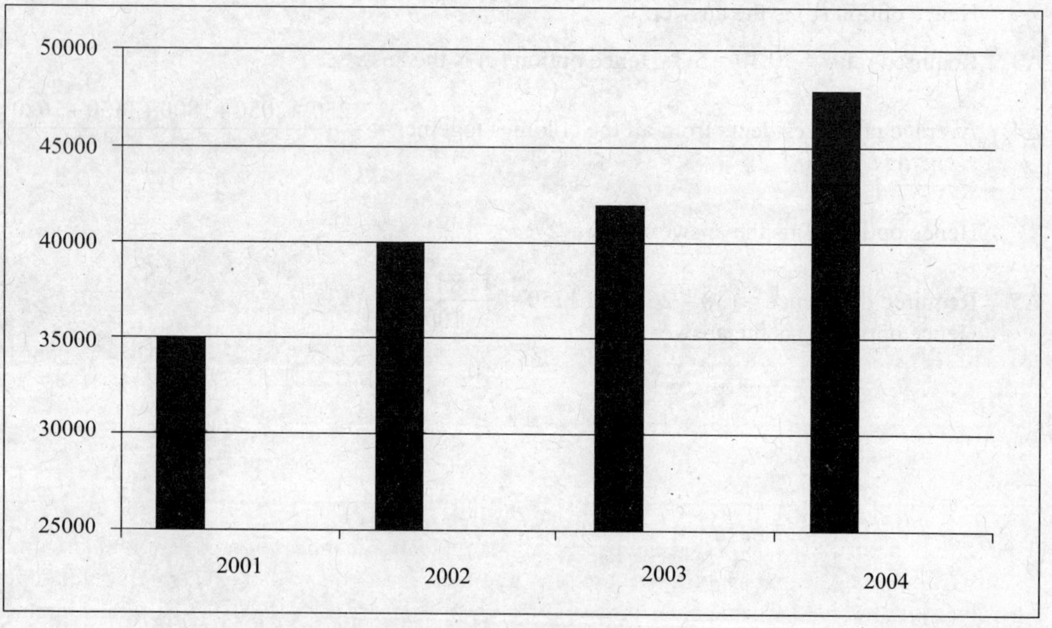

Answer all the questions based upon the pie charts and bar chart given above.

Q1. In which faculty there was decrease in the number of students from 2001 to 2002?
(a) None (b) Commerce (c) Agriculture (d) Pharmacy (e) None of these

Q2. What is the ratio between the number of students studying pharmacy in the years 2001 and 2002 respectively?
(a) 4:3 (b) 41:72 (c) 2:3 (d) 49:72 (e) None of these

Q3. What was the approximate percentage increase in the number of students of Engineering from the year 2001 to 2002?
(a) 17% (b) 13% (c) 5% (d) 23% (e) 9%

Q4. In the year 2001, the number of student studying Arts and Commerce together is what percentage of the number of students studying these subjects together in 2002?
(a) 76 (b) 92 (c) 82 (d) 79 (e) 97

Q5. In which of the following faculties, the percentage change (consider absolute value) in the number of students was minimum from 2001 to 2002?
(a) Arts (b) Science (c) Commerce (d) Medicine (e) Engineering

Rough Work

Answer Grid

Q1	Q2	Q3	Q4	Q5
c	d	e	b	a

STORYLINE

There are two pie charts depicting the percentage composition/contribution of different streams/faculty from a state. Bar chart provides the total number of students in those two years (including other two years). There are questions that require you to compare the values of a particular stream across the two years. You may expect pie chart based percentage calculations in the whole set.

Solutions to Q1–Q5

A1. Let us first spend some time observing the data set, and identify if we can eliminate the some options through observation without doing actual calculation:
First, we will observe the data for options (b), (c) and (d).
Looking at option (d), Pharmacy cannot have decline in the data as its percentage share is rising and total number of students is also rising. Similarly, option (b) cannot be the answer as share of commerce is fixed at 24%.
Number of students decreased from 2800 to 2000 in Agriculture. Hence option (c) is the answer.

A2. Number of students studying Pharmacy in 2001 = 2450
Number of students studying Pharmacy in 2002 = 3600
Therefore, required ratio = 2450 : 3600 = 49 : 72
Hence option (d) is the answer.

A3. Difference= 7600 – 7000 = 600

Hence required percentage increase = $\frac{600}{7000} \times 100 = 8.57\%$

Hence option (e) is the answer.

A4. Total number of students in Arts and Commerce faculties in 2001 = 4200 + 7700 = 11900
Total number of students in Arts and Commerce faculties in 2002 = 4400 + 9600 = 1400

Therefore required percentage= $\frac{12950}{14000} \times 100 = 92\%$

Hence option (b) is the answer.

A5. Question is asking for absolute value in the percentage change of different faculty across the two years ⇒ So we are required to calculate percentage increase as well as percentage decrease both, as applicable for different faculty.
Again, let us first observe the data set and identify if we can eliminate any option by simple observation of data:
We can eliminate option (b) Science – Percentage contribution of science has increased from 16% to 22% - so it cannot be the answer. (Going through the actual calculation,

Science = $\frac{8800 - 5600}{5600} \times 100 = 57\%$)

Similarly, option (c) Commerce cannot be the answer as it will have net percentage change = 14.28%. (We will calculate the change in total number of students only, and since the percentage contribution of commerce remains same in both the years, answer would be equivalent to the percentage change in the total number of students.) Hence even option (c) cannot be the answer.
Let us calculate for remaining three options:
Percentage change in the faculty of:

Arts = $\frac{4550 - 4400}{4400} \times 100 = 3.2\%$

$$\text{Engineering} = \frac{7600 - 7000}{7000} \times 100 = 8.57\%$$

$$\text{Medicine} = \frac{4200 - 4000}{4200} \times 100 = 4.7\%$$

[Since question asks for absolute value of percentage change, we do not care for percentage increase or percentage decrease].
Hence, the percentage change in Arts is minimum.
Hence option (a) is the answer.

IPL

Set 58

Direction for questions 1–5

There were 7 teams—A, B, C, D, E, F and G, during the last IPL. Each team played three matches with every other team. All the matches resulted in win—loss situation i.e., none of the matches played were a tie.

To decide the winner, if was decided to calculate a ratio called as Win Ratio.

Win ratio = Number of matches won/number of matches played

Following table gives the outcome of some of the matches played in the series:

	A	B	C	D	E	F	G	Win ratio
A	NA	2		3	1			0.61111...
B		NA	1	0		0	1	
C	2		NA		3	2		0.7777...
D			1	NA	0		2	0.389
E		2			NA	2		
F	1					NA	1	0.5
G			0		3		NA	

So, team C won 2 matches against team A and so on. As it can be seen, some of the data in the above table are missing because of some problem in the computer.

Q1. How many matches did team C won against team B?
(a) 0 (b) 1 (c) 2 (d) 3 (e) Cannot be determined

Q2. Which team won the maximum number of matches?
(a) A (b) B (c) C (d) E (e) F

Q3. For how many matches, the unique winner team cannot be decided?
(a) 0 (b) 1 (c) 2 (d) 3 (e) 5

Q4. What is the total number of matches played in the whole series?
(a) 18 (b) 21 (c) 63 (d) 30 (e) None of these

Q5. What is the total number of matches won by team E?
(a) 8 (b) 9 (c) 10 (d) 11 (e) None of these

Rough Work

Answer Grid

Q1	Q2	Q3	Q4	Q5
c	c	a	c	b

STORYLINE

There are seven teams, and each team is playing three matches with every other team. Question provides the data in the format of the ratio Win: Total matches. You are required to convert the win ratio into number of matches won. For some of the teams, win ratio is not given. In those cases, you may be required to count the total number of wins, and then proceed. Between any pair of teams, totan matches played = total number of wins = total number of losses = 3.

Solutions to Q1–Q5

Number of matches played by each team against other team = 3 matches.
Total number of matches played by each team = 3 matches × 6 teams = 18 matches.

$$\text{Win ratio} = \frac{\text{Number of matches won}}{\text{Number of matches played}}$$

So Number of matches won = Win ratio × Number of matches played (=18)

For Team A:

Win ratio $= 0.6111\ldots = \frac{11}{18} \Rightarrow$ So team A won 11 matches out of total 18 matches played.

For Team C:

Win ratio $= 0.7777\ldots = \frac{14}{18} \Rightarrow$ So team C won 14 matches out of total 18 matches played.

For Team D:

Win ratio $= 0.389\ldots = \frac{7}{18} \Rightarrow$ So team C won 7 matches out of total 18 matches played.

For Team F:

Win ratio $= 0.5 = \frac{9}{18} \Rightarrow$ So team F won 9 matches out of total 18 matches played.

Total number of win = Total number of loss = Total number of matches $= \frac{18 \times 7}{2} = 63$ matches

(One team plays 18 matches, so all the seven teams play 18 × 7 = 126 matches. Besides, all the matches have been counted twice. So unique number of matches $= \frac{126}{2} = 63$ matches.

Alternatively, to calculate the total number of matches, we know that the total number of matches in a round robin league tournament $= {}^{n}C_{2}$. Since every team is playing 3 matches with every other team, total number of matches $= {}^{n}C_{2} \times 3$. In this case, $n = 7$. Hence total number of matches $= {}^{7}C_{2} \times 3 = 63$ matches.)

Let us fill the table now:

Original table

	A	B	C	D	E	F	G	Win ratio
A	NA	2		3	1			0.61111
B		NA	1	0		0	1	
C	2		NA		3	2		0.7777
D			1	NA	0		2	0.389
E		2			NA	2		
F	1					NA	1	0.5
G			0		3		NA	

First of all, we transpose the values. For example:

	A	B
A	NA	2

Team A won 2 matches against team B, so team B won the remaining 1 match against team A. And so on for other teams. Once we have got these individual values through transposing the data, to find out any remaining value, we can simply subtract the matches won from total number of matches won (as calculated using win ratio). So Team A won 11 – (2 + 1 + 3 + 1 + 2) = 2 matches against team G ⇒ Team G won 1 match against team A.
Original table

	A	B	C	D	E	F	G	Win ratio	No. of wins
A	NA	2	1	3	1	2	2	0.61111	11
B	1	NA	1	0	1	0	1	4/18 = 0.22...	4
C	2	2	NA	2	3	2	3	0.7777	14
D	0	3	1	NA	0	1	2	0.389	7
E	2	2	0	3	NA	2	0	9/18 = 0.5	9
F	1	3	1	2	1	NA	1	0.5	9
G	1	2	0	1	3	2	NA	9/18 = 0.5	9

Now all the questions can be answered:

A1. Team C won 2 matches against team B. Hence option (c) is the answer.

A2. Team C wins maximum number of matches. Hence option (c) is the answer.

A3. Winner can be decided for every match. Hence option (a) is the answer.

A4. Total number of matches played = 63. Hence option (c) is the answer.

A5. Total number of matches won by team E = 9. Hence option (b) is the answer.

Plastic Money Spending Set 59

Direction for questions 1–3

The Reserve Bank of India has reduced transaction charges for payments made online through debit cards to a maximum of 1%, addressing a long-standing grouse of consumers and merchants who were paying almost double earlier. Further, each of the debit card issuing agencies has been directed to issue debit card for a minimum of five years.

From 1 July 2011, transaction fee for online payments up to Rs 2000 made through debit cards has been slashed to 0.75% of the value of the goods or services bought. The charge can be a maximum of 1% for purchases above that value. Banks were earlier charging 1.8% of all online payments made using debit cards.

The move is expected to boost e-commerce in India, Asia's third-largest economy, where use of internet for purchase of goods and services is growing at 30%–35% every year.

According to the Internet and Mobile Association of India (IAMAI), e-transactions in the country totaled 46000 crore in 2011.

Over the past five years, the value of debit card transactions has risen 355% to 26418.10 crore in the year 2011, according to RBI. However, more than 95% of this is offline transaction.

Over the same period, debit card usage, both in terms of number of transactions (primarily offline) and growth in value, has outstripped credit card spending by a huge margin. The number of debit cards issued has risen by 265% or by 18.2 crore in 2011 while the number of credit cards have increased by just 5.5% to 1.82 crore in 2011.

Usage of debit cards is also higher than that of credit cards in terms of total number of transactions. Between 2005-06 and 2010–11 (provisional), the number of transactions for debit cards grew 270% to 17.02 crore. During the same period, the number of credit card transactions increased by 50% to 23.4 crore.

The RBI's directive to banks on online transactions was issued under Section 18 of the Payment and Settlement Systems Act, 2007 (Act 51 of 2007). Banks don't charge a fee for offline use of debit cards.

Q1. In the year 2011, what is the value of online debit card transaction in India (in Rs Crores)?

(a) 1320 (b) 1246 (c) 1576 (d) Cannot be determined

Q2. What is the average spending per debit card in the year 2011?

(a) Rs 1451 (b) Rs 14510 (c) Rs 145.1 (d) Cannot be determined

Q3. Mr M Shashi Kumar has been avid user of ejay.com, an online portal for buying various electronic goods. He recalls the total value of online goods that he purchased in the years using his debit card online:

Timeframe	Spending (in Rs)
On or After 1 July 2011	1500
Before 1 July 2011	22000
2010	17700
2009	20800
2008	14500

What is the total transaction fee paid by Mr M Shashi Kumar on the spending given above?
(a) Rs 13561.25 (b) Rs 1426.25 (c) Rs 1361.25 (d) Cannot be determined

Rough Work

Answer Grid

Q1	Q2	Q3
a	d	c

Solutions to Q1–Q3

A1. Value of debit card transactions in 2011 = Rs 26418.10 crore.
Online transaction = 5% of 26418.10 crore = Rs 1320 crores. Hence option (a) is the answer.

A2. Though the total debit card spending in 2011 is given, passage does not provide the total number of active debit cards in the year 2011. It only states that in 2011, total number of new debit cards issued = 18.2 crores. Hence average spending per debit card cannot be determined. Hence option (d) is the answer.

A3. This is mentioned that:
"From 1 July 2011, transaction fee for online payments up to Rs 2000 made through debit cards has been slashed to 0.75% of the value of the goods or services bought. The charge can be a maximum of 1% for purchases above that value. Banks were earlier charging 1.8% of all online payments made using debit cards."
All the online debit card transactions before 1 July 2011 used to attract a flat transaction fee of 1.8%. So instead of calculating the transaction fee for each of the years, we simply add the total buying before 1 July 2011, and calculate 1.8% of that.

Before 1st July 2011	22000
2010	17700
2009	20800
2008	14500
Sum	75000

So, total transaction fees paid by him before 1 July 2011 = 1.8% of 75000 = Rs 1350
Post 1 July 2011, Total transaction value = Rs 1500. Hence transaction fee = 0.75% of Rs 1500 = Rs 11.25.
Hence total transaction fees paid = Rs 1350 + Rs 11.25 = Rs 1361.25. Hence option (c) is the answer.

Auto and Taxi Services Set 60

Direction for questions 1–4

Read the passage below and solve the questions based on it.

Number of taxi drivers and auto drivers working in three shifts are given below:

Vehicle	Shift	Number of drivers	Working hours	Total collection (in Rs thousands)
Taxi	M1	425	10	132.1
	M2	672	8	240.2
	M3	135	6	42.3
Auto	M1	350	9	57.3
	M2	425	8	100.8
	M3	95	7	18.9

A taxi driver charges Rs 13 for every 1.6 km, while an auto driver charges Rs 8 for every 1.3 km. For the night shift, charges are 1.5 times the normal rate. In case of Auto, M3 is the night shift.

Q1. By what percentage is the average collection per taxi driver more during the M2 shift than during the M1 shift?

(a) 4% (b) 15% (c) 34% (d) 82%

Q2. What is the average distance traveled by the auto drivers during the M3 shift?

(a) 1356 km (b) 2047 km (c) 3054 km (d) 2672 km

Q3. Who get the highest returns for their efforts with respect to work hours?

(a) Auto Drivers – M3 Shift (b) Taxi drivers – M2 shift
(c) Taxi drivers – M3 shift (d) Auto drivers – M1 shift

Q4. Which of the following is false?

(a) Lowest contribution per driver is by auto drivers in the M1 shift.
(b) Lowest returns for the efforts are for auto drivers in the M1 shift.
(c) The percentage of drivers during the M3 shift is 21% of the total drivers.
(d) None of these

Rough Work

Answer Grid

Q1	Q2	Q3	Q4
b	b	c	c

STORYLINE

Data set provides data of Taxi and Auto services, number of drivers, working hours and total collection by them. Further, out of three shifts, one shift is night shift and gets paid more.

Solutions to Q1–Q4

In case of Taxi:

Per person in Rs — Per hour in Rs

M1 = $\frac{132.1}{425} = 310.82$ — $\frac{310.82}{10} = 31.08$

M2 = $\frac{240.2}{672} = 357.44$ — $\frac{357.44}{8} = 44.60$

M3 = $\frac{42.3}{135} = 313.33$ — $\frac{313.33}{6} = 52.22$

In case of Auto:

Per person in Rs — Per hour in Rs

M1 = $\frac{57.3}{350} = 163.714$ — $\frac{163.714}{9} = 18.19$

M2 = $\frac{100.8}{425} = 237.176$ — $\frac{237.176}{8} = 29.647$

M3 = $\frac{18.9}{95} = 198.947$ — $\frac{198.947}{7} = 28.421$

A1. Data can be noted down from the above calculation:
Average collection per taxi driver in M2 shift = 357 (approx)
Average collection per taxi driver in M1 shift = 310 (approx)
So, percentage increase $= \frac{357.44 - 310.82}{310.82} = 15\%$
Hence option (b) is the answer.

A2. M3 shift is night shift in case of Auto.
Total distance travelled $= \frac{18900 \times 1.3}{8} \times \frac{2}{3} = 2047$ km
Hence option (b) is the answer.

A3. It can be seen from the data calculated above that Taxi drivers earn maximum in M3 shifts compared to all the other shifts in Auto/Taxi.
Hence option (c) is the answer.

A4. M3 = 135 + 95 = 230
Total drivers = 1872 + 230 = 2102
Therefore $\frac{230}{2102} \times 100 =$ less than 10%
Hence option (c) is the answer.

True Lies

Set 61

Direction for questions 1–6

The Financial Statements of Neo Corporation for the period 2004–2008 are summed up given in the table below.

Particulars	2004	2005	2006	2007	2008
Revenue	10,156.00	11,799.00	14,380.00	17,996.00	22,430.00
COGS	1,770.00	2,033.00	2,516.0	3,349.00	3,984.00
Gross Profit	8,386.00	9,766.00	11,864.00	14,647.00	18,446.00
Net Income	2,681.00	2,886.00	3,381.00	4,274.00	5,521.00
Total Assets	12,763.00	20,687.00	29,029.00	34,572.00	47,268.00
Total Current Liabilities	4,272.00	8,063.00	6,930.00	9,387.00	10,029.00
Total Liabilities	4,768.00	9,850.00	14,017.00	17,653.00	24,243.00
Total Equity	7,995.00	10,837.00	15,012.00	16,919.00	23,025.00
Total Liabilities and Equity	12,763.00	20,687.00	29,029.00	34,572.00	47,268.00

Evaluate the pairs of statements in each question on the basis of data given in the table and mark your answer as option:

(a) if only statement I is correct.
(b) if only Statement II is correct.
(c) if both the statements are correct.
(d) if none of the statements is correct.

Q1. Statement I: Percentage Revenue growth over the previous year has been the highest in the year 2007.
Statement II: Percentage Revenue growth over the previous year has been the lowest in the year 2008.

Q2. Statement I: Lowest growth rate over the previous year in Total Assets was recorded in the year 2006.
Statement II: Highest growth rate over the previous year in Total Equity was recorded in the year 2008.

Q3. Statement I: Gross Profit every year has recorded higher growth rate as compared to the previous year for the period under review.
Statement II: Total Assets every year has not recorded higher growth rate as compared to the previous year for the period under review.

Q4. Statement I: The COGS to Revenue ratio is the lowest in the year 2006.
Statement II: The COGS to Revenue ratio is the highest in the year 2007.

Q5. Statement I: The Gross Profit to Revenue ratio is the lowest in the year 2007.
Statement II: The Gross Profit to Revenue ratio is the highest in the year 2005.

Q6. Statement I: Total Assets to Total Liabilities Ratio is the lowest in the year 2007.
Statement II: Total Assets to Total Liabilities Ratio is the highest in the year 2005.

Rough Work

Answer Grid

Q1	Q2	Q3	Q4	Q5	Q6
a	d	d	c	c	d

Solutions to Q1–Q6

A1. Percentage Revenue growth is as follows:

$$2005 = \frac{11799 - 10156}{10156} \times 100 = 16.178\%$$

$$2006 = \frac{14380 - 11799}{11799} \times 100 = 21.875\%$$

$$2007 = \frac{11796 - 14380}{14380} \times 100 = 25.146\%$$

$$2008 = \frac{22430 - 17996}{17996} \times 100 = 24.639\%$$

Percentage Revenue growth over the previous year has been the highest in the year 2007.
⇒ Statement I is correct.
Percentage Revenue growth over the previous year has been the lowest in the year 2005.
⇒ Statement II is incorrect.
So, option (a) is the answer.

A2. Total Assets growth rate is as follows:

$$2005 = \frac{20687 - 12763}{12763} \times 100 = 62.086\%$$

$$2006 = \frac{29029 - 20687}{20687} \times 100 = 40.325\%$$

$$2007 = \frac{34572 - 29029}{29029} \times 100 = 19.095\%$$

$$2008 = \frac{47268 - 34572}{34572} \times 100 = 36.723\%$$

Lowest growth rate over the previous year in Total Assets was recorded in the year 2007.

⇒ Statement I is incorrect.

Total Equity growth rate is as follows:

$$2005 = \frac{10837 - 7995}{7995} \times 100 = 35.547\%$$

$$2006 = \frac{15012 - 10837}{10837} \times 100 = 38.525\%$$

$$2007 = \frac{15012 - 10837}{10837} \times 100 = 12.703\%$$

$$2008 = \frac{16919 - 15012}{15012} \times 100 = 36.089\%$$

Highest growth rate over the previous year in Total Equity was recorded in the year 2006.

⇒ Statement II is incorrect.

So, option (d) is the answer.

A3. Gross Profit growth rate is as follows:

$$2005 = \frac{2766 - 8386}{8386} \times 100 = 16.456\%$$

$$2006 = \frac{11864 - 9766}{9766} \times 100 = 21.483\%$$

$$2007 = \frac{14647 - 11864}{11864} \times 100 = 23.458\%$$

$$2008 = \frac{18446 - 14647}{14647} \times 100 = 25.937\%$$

Gross Profit every year has recorded higher growth rate as compared to the previous year for the period under review.

⇒ Statement I is correct.

From the solution to the previous question, we can conclude that:

Total Assets every year has not recorded higher growth rate as compared to the previous year for the period under review.

⇒ Statement II is correct.

So, option (c) is the answer.

A4. COGS to Revenue ratio is as follows:

$$2004 = \frac{1770}{10156} = 0.1743$$

$$2005 = \frac{2033}{11799} = 0.1723$$

$$2006 = \frac{2516}{14380} = 0.1750$$

$$2007 = \frac{3984}{22430} = 0.1861$$

$$2008 = \frac{8386}{10156} = 0.1776$$

The COGS to Revenue ratio is the lowest in the year 2005.
⇒ Statement I is incorrect.
The COGS to Revenue ratio is the highest in the year 2007.
⇒ Statement II is correct.
So, option (b) is the answer.

A5. The Gross Profit to Revenue ratio is as follows:

$$2004 = \frac{8386}{10156} = 0.8257$$

$$2005 = \frac{9766}{11799} = 0.8277$$

$$2006 = \frac{11864}{14380} = 0.8250$$

$$2007 = \frac{14647}{17996} = 0.8139$$

$$2008 = \frac{18446}{22430} = 0.8224$$

The Gross Profit to Revenue ratio is the lowest in the year 2007.
⇒ Statement I is correct
The Gross Profit to Revenue ratio is the highest in the year 2005.
⇒ Statement II is correct.
So, option (c) is the answer.

A6. Total Assets to Total Liabilities Ratio is as follows:

$$2004 = \frac{12763}{4768} = 2.677$$

$$2005 = \frac{20687}{9850} = 2.100$$

$$2006 = \frac{29029}{14017} = 2.071$$

$$2007 = \frac{34572}{17653} = 1.958$$

$$2008 = \frac{47268}{24243} = 1.950$$

Total Assets to Total Liabilities Ratio is the lowest in the year 2008.
⇒ Statement I is incorrect.
Total Assets to Total Liabilities Ratio is the highest in the year 2004.
⇒ Statement II is incorrect.
So, option (d) is the answer.

Climate Control

Set 62

Direction for questions 1–4

Study the following graph and answer the questions that follows:

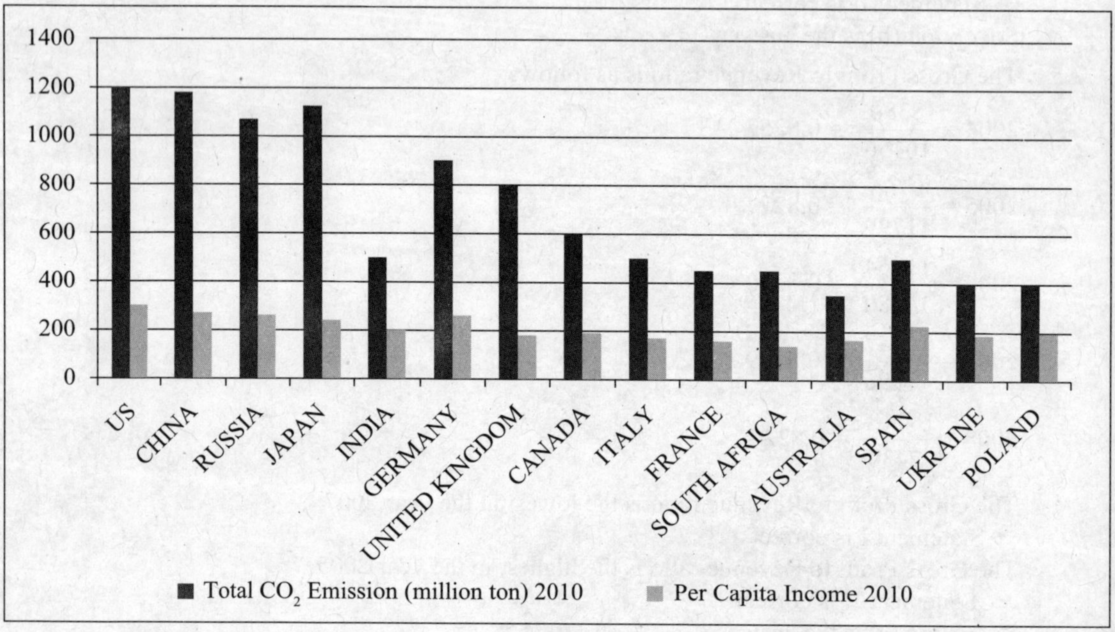

Q1. If the world energy council formulates a norm for high emission countries to reduce their emission each year by 12.5% for the next two years then what would be the ratio of CO_2 emission to per capita income of US, China and Japan after two years. The per capita income of China, Japan and US is expected to increase every year by 4%, 3% and 2% respectively.

(a) 3.5, 3.8, 401

(b) 3.4, 3.5, 3.9

(c) 2.9, 3.1, 3.4

(d) None

Q2. If US and China, decide to buy carbon credits, from Spain and Ukraine to make up for their high emissions, then in how many years US, and China would be able to bring down its ratio of CO_2 *emission (million ton) to per capita income* to world standard benchmark of 0.75. (per capita income of the given countries remain same, 0.5 CO_2 emissions [million ton] is compensated by purchase of 1.25 units of carbon credit, and a country can buy carbon credit units in three lots of 15, 20 and 30 units in a single year.)

(a) 3.8 years (b) 38 years

(c) 30 years (d) None of these

Q3. France, South Africa, Australia, Ukraine and Poland form an energy consortium which declares CO_2 emission of 350 million ton per annum as standard benchmark. The energy consortium decides to sell their carbon emission savings against the standard benchmark to high carbon emission countries. It is expected that the per capita income of each country of the energy consortium increases by 2%, 2.5% and 3.5% p.a. for the next three years respectively. The ratio *of CO_2 emission to per capita income* of the each energy consortium country reduces by 50% and remains constant for the next three years. By selling 0.5 CO_2 emissions [million ton] the energy consortium earns 1.25 carbon credits, then determine the total carbon credits earned by energy consortium in three years.

(a) 3560 (b) 4506

(c) 5060 (d) None

Q4. Select the wrong statement in reference to the position of India vis-à-vis other countries in the graph in terms of the ratio of CO_2 emission to per capita income increasing order).

(a) India stands at 5th position if 50 is added to the given per capita income figures of each country.

(b) India stands at 5th position at the given CO_2 emission level and per capita income of each country.

(c) India stands at 5th position if 200 million ton CO_2 emission is deducted from the given CO_2 emission figures of each country.

(d) India stands at 5th position if 200 million ton CO_2 emission is deducted from the given CO_2 emission figures of each country and 50 is added to the given per capita income of each country.

Rough Work

Answer Grid

Q1	Q2	Q3	Q4
c	b	c	d

Solutions to Q1–Q4

A1. The ratio of CO_2 emission to per capita income for USA, China and Japan after two years will be:

$$\text{USA} = \frac{1200 \times 0.875 \times 0.875}{300 \times 1.02 \times 1.02} = 2.9$$

$$\text{China} = \frac{1180 \times 0.875 \times 0.875}{270 \times 1.04 \times 1.04}$$

$$\text{Japan} = \frac{1127 \times 0.875 \times 0.875}{240 \times 1.03 \times 1.03}$$

Hence, option (c) is the answer.

A2. To bring down the ratio of CO_2 *emission to per capita income* to meet the standard benchmark of 0.75, US will have to bring down their CO_2 emission to 300 × 0.75 = 225 and China will have to bring down its emission to 270 × 0.75 = 202.5 million tons.
So, the quantity of CO_2 emission cut by US = 1200 – 225 = 975 million tons and the quantity of CO_2 emission cut by China = 1180 – 202.5 = 977.5 million tons.
It is given that:
To reduce 0.5 million tons of CO_2 emission, a country has to purchase 1.25 units of carbon credits ⇒ To reduce 1 million tons of CO_2 emission 2.5 units of carbon credits need to be purchased.
Hence, to reduce CO_2 emission by 975 and 977.5 million tons, the quantity of carbon credits that need to be purchased by the US and China are (975 × 2.5) and (977.5 × 2.5) i.e. 2437.5 and 2443.75 respectively.
Number of carbon credits that a country can buy in a year = 65 (=15 + 20 + 20).

So, the number of years that U.S. will take to get the ratio to the required level = $\frac{2437.5}{65}$ = 37.5 years. Same data for China will be = $\frac{2443.75}{65}$ = 37.6 years

So, in 38 years US and China will be able to bring down its CO_2 emission to per capita income to the world standard benchmark of 0.75.
Hence, option (b) is the answer.

A3. By selling 0.5 CO_2 emissions the energy consortium earns 1.25 carbon credits.
Hence number of carbon credits earned by any country in the consortium = (Total CO_2 emission saved) × (1.25/0.5)
Total CO_2 emission saved by any country in the consortium = 350 – revised CO_2 emission by that country.
Revised total CO_2 emission per year for each country in the consortium = Revised ratio × revised per capita income per year. Now calculations can be done.
Hence, option (c) is the answer.

A4. Using option (b), following table can be made:

Country	CO_2 emission/Per capita income
US	4
China	4.37
Russia	4.12
Japan	4.7
India	2.5
Germany	3.46
UK	4.44
Canada	3.16
Italy	2.94
France	2.81
SA	3.21
Australia	2.18
Spain	2.22
Ukraine	2.16
Poland	2.05

From the above table, we can conclude that India stands at the 5th position at the given CO_2 emission level and per capita income of each country (in increasing order).
Hence, option (b) is true.
Consider option (a):
Since 50 is added to the per capita income for each country, the ratio given in the above table decreases for each country and for India it becomes 2.
Only four countries (Australia, Spain, Ukraine and Poland) have the required ratio less than India.
Hence, option (a) is also true.
Consider option (c):
Since 200 million tons are deducted from the given CO_2 emission value, the ratio given in the above table decreases for each country and for India it becomes 1.5. $\Rightarrow$ Any country with CO_2 emission value > 1.5 times of revised per capita income will have a ratio greater than India.
Only four countries (Australia, Spain, Ukraine and Poland) have the required ratio less than India.
Hence India stands at the 5th position in this case also.
Hence, the statement in option (c) is also true.
So, options (a), (b) and (c) are true.
Hence, option (d) is the answer.

Asset Utilization on CSR Set 63

Direction for questions 1–4

Go through the charts given below and solve the questions based on it.

The following graphs give annual data of *Assets, Sales (as percentage of Assets) and Spending on Corporate Social Responsibility* (CSR) (as percentage of Sales), of a company for the period 2004–2009.

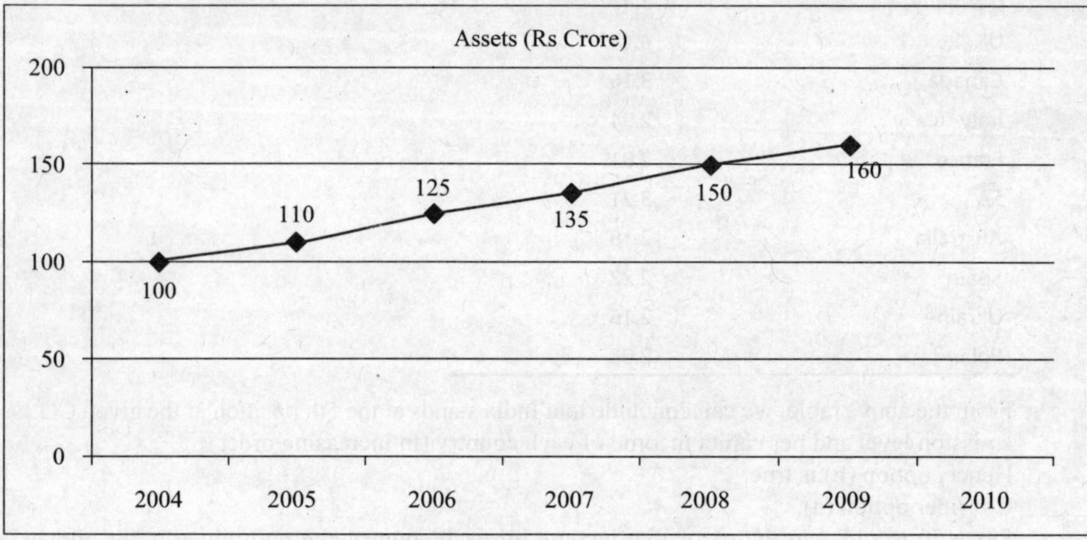

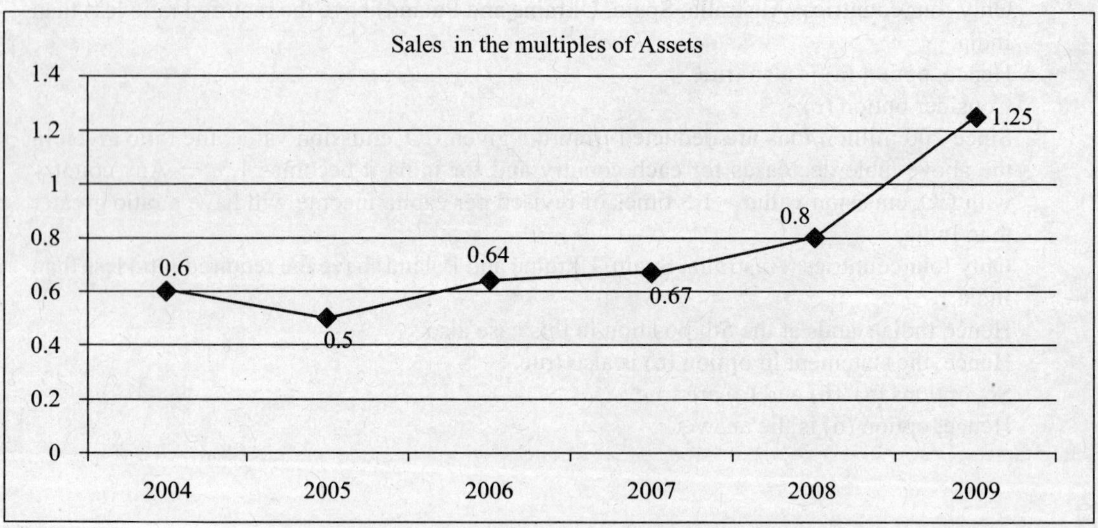

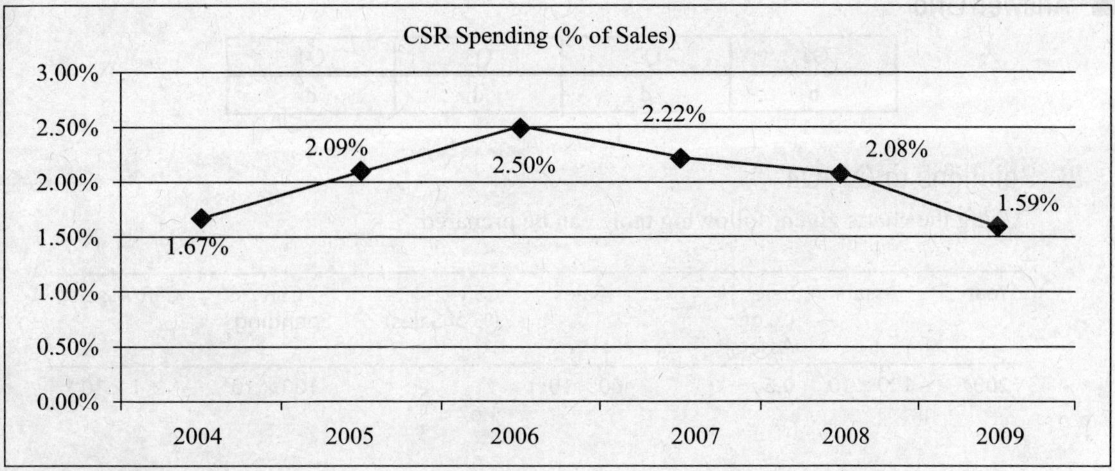

Q1. In which year was the increase in spending on CSR, vis-à-vis the previous year, the maximum?
(a) 2005 (b) 2006 (c) 2007 (d) 2008

Q2. Of the years indicated below, in which year was the ratio of CSR/Assets the maximum?
(a) 2004 (b) 2005 (c) 2006 (d) 2008

Q3. What was the maximum value of spending on CSR activities in the period 2004–2009?
(a) Rs 0.5 Crore (b) Rs 1.0 Crore (c) Rs 2.0 Crore (d) Rs 3.0 Crore

Q4. In which year, did the spending on CSR (measured in Rs) decline, as compared to previous year?
(a) 2006 (b) 2007 (c) 2008 (d) None of above

Rough Work

Answer Grid

Q1	Q2	Q3	Q4
b	d	d	d

Solutions to Q1–Q4

Using the charts given, following table can be prepared:

Year	Assets	Sales (% of Assets)	Sales	CSR (% of Sales)	CSR Spending	CSR/Assets
2004	100×10^7	0.6	60×10^5	$1.67 = \frac{5}{3}$	100×10^3	1×10^{-4}
2005	100×10^7	0.5	55×10^5	2.09	114.95×10^3	1.045×10^{-4}
2006	125×10^7	$0.64 = \frac{16}{25}$	80×10^5	$2.5 = \frac{5}{2}$	200×10^3	1.6×10^{-4}
2007	135×10^7	$0.67 = \frac{2}{3}$	90×10^5	$2.22 = \frac{20}{9}$	200×10^3	1.48×10^{-4}
2008	150×10^7	$0.8 = \frac{8}{10}$	120×10^5	2.06	247.2×10^3	1.648×10^{-4}
2009	160×10^7	$1.25 = \frac{5}{4}$	200×10^5	1.58	316×10^3	1.975×10^{-4}

Now all the questions can be answered.

A1. Increase in spending on CSR was maximum in 2006.
Hence, option (b) is the answer.

A2. The ratio of CSR/Assets was the maximum in 2008 in the years given in the options.
Hence, option (d) is the answer.

A3. Maximum spending on CSR was approx. Rs 3 Crores.
Hence, option (d) is the answer.

A4. CSR does not decrease during the specified period.
Hence, option (d) is the answer.

Production by Four Companies in Six Years

Set 64

Direction for questions 1–5

Answer the questions based on the following graph.

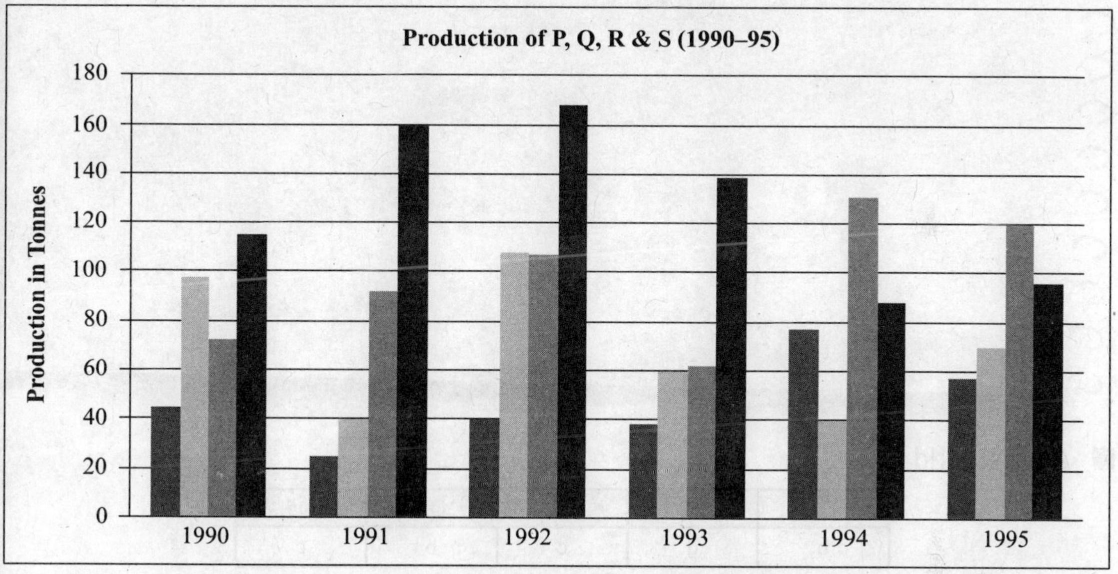

Q1. In which year the growth rate of total production (of all products) is highest over the previous year?

(a) 1991 (b) 1992 (c) 1993 (d) 1995

Q2. If the stability of the production during 1990 to 1995 is defined as,

$$\frac{\textit{Average Production}}{\textit{Maximum Production} - \textit{Minimum Production}}$$ then which product is most stable?

(a) Product P (b) Product Q (c) Product R (d) Product S

Q3. If four products P, Q, R and S shown in the graph are sold at price of Rs 9, Rs 4, Rs 13 and Rs 3 respectively during 1990–1995, then the total revenue of all the products is lowest in which year?

(a) 1991 (b) 1992 (c) 1993 (d) None of the above

Q4. Individual revenue of P, Q, R and S for the entire period (1990–1995) is calculated based on the price of Rs 9, Rs 4, Rs 13 and Rs 3 respectively. Which product fetches the lowest revenue?

(a) Product P (b) Product O (c) Product R (d) Product S

Q5. Four products P, Q, R and S shown in the graph are sold at price of Rs 9, Rs 4, Rs 13 and Rs 3 respectively during 1990–1995. Which of the following statements is TRUE?
(a) Product R fetches second highest revenue across products in 1991.
(b) Sum of revenue of P, Q and S is more than the revenue of R in 1994.
(c) Cumulative revenue of P and Q is more than the revenue of S in 1993.
(d) None of the above

Rough Work

Answer Grid

Q1	Q2	Q3	Q4	Q5
b	d	c	b	c

Solutions to Q1–Q4

A1. Following is the table presenting the total production across the years:

Year	Sum of Production
1990	330
1991	316
1992	423
1993	299
1994	336
1995	343

It can be clearly seen that in the year 1992, total production is highest over the previous year's total production. Hence option (b) is the answer.

A2. Following table can be obtained:

Company	P	Q	R	S
Average	47.4	63.6	102.4	130
Diff b/w max and min	52	68	69	80
Stability	0.91	0.93	1.48	1.62

Hence option (d) is the answer.

A3. Following table can be obtained:

Year	P	Q	R	S	Summation
1990	405	392	936	345	2078
1991	225	160	1196	477	2058
1992	360	432	1391	504	2687
1993	342	240	806	417	1805
1994	693	160	1703	264	2820
1995	513	280	1560	288	2641

It can be seen that the minimum revenue is obtained in the year 1993. Hence option (c) is the answer.

A4. Following table can be obtained:

	P	Q	R	S
Total production	282	416	584	765
Unit Price (Rs/ unit)	9	4	13	3
Total revenue (Rs)	2538	1664	7592	2295

Hence option (b) is the answer.

Population and Literacy Rate

Set 65

Direction for questions 1–3

The following graph shows population data (males and females), educated people data (males and females) and number of male in the population for a given period of 1995 to 2010. All data is in million.

From the information given in the graph answer the questions that follow

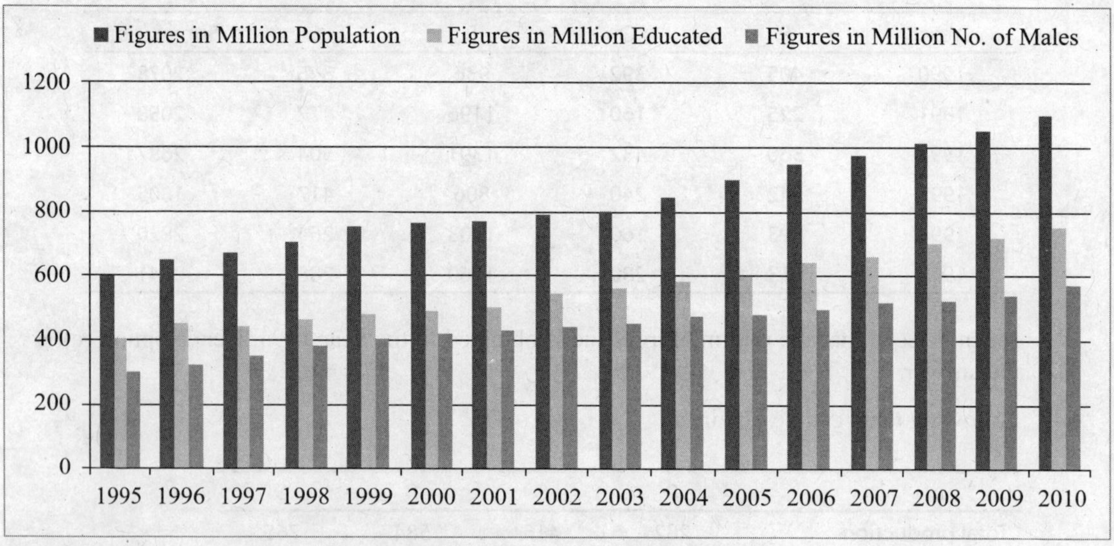

Q1. In which year the percentage increase in the number of females over the previous year is highest?

(a) 1996 (b) 1999 (c) 2004 (d) 2005

Q2. In 2002 if the ratio of number of educated male to professionally educated female was 5:4. If the number of educated males increased by 25% in 2003. What is the percentage change in number of uneducated females in 2003?

(a) +25% (b) +30% (c) +34% (d) +56%

Q3. In year 2005 total population living in urban area is equal to sixty eight per cent of educated population. The ratio of number of people living in urban area to people living in rural area is 43:12 in 2010. What is the ratio of the rural population in 2005 to that in 2010?

(a) 0.8 (b) 0.47 (c) 2.05 (d) None

Rough Work

Answer Grid

Q1	Q2	Q3
d	d	c

Solutions to Q2–Q3

A2. No. of educated people in 2002 = 545
Therefore, no. of educated males in 2002 = (5/9) × 545 ≈ 303
No. of educated females in 2002 = 545 – 303 = 242
Now, no. of females in 2002 = Total population – Total male population = 350
So, no. of uneducated females in 2002 = 350 – 242 = 108
The no. of educated males increases by 25% in 2003.
Hence, no. of educated males in 2003 = 1.25 × 303 ≈ 379
So, no. of educated females in 2003 = 560 – 379 ≈ 181
Now, no. of females in 2003 = Total population – Total male population = 350
So, no. of uneducated females in 2003 = 350 – 181 ≈ 169

Hence, percentage change in uneducated females $\frac{169-108}{108} \times 100 \approx 56\%$
Hence, option (d) is the answer.

A3. Total population of urban area in 2005 = 0.68 × 600 = 408 million
Total population in 2005 = 900 million
Hence, population of rural area in 2005 = 900 – 408 = 492 million

Population of rural area in 2010 = $\frac{12}{55} \times 1100 = 240$ million
Therefore, the required ratio = 492/240 = 2.05
Hence, option (c) is the answer.

Road Network

Set 66

Direction for questions 1–5

Read the information given below and solve the questions based on it.

A significant amount of traffic flows from point S to point T in the one-way street network shown below.

Points A, B, C, and D are junctions in the network, and the arrows mark the direction of traffic flow. The fuel cost in rupees for travelling along a street is indicated by the number adjacent to the arrow representing the street.

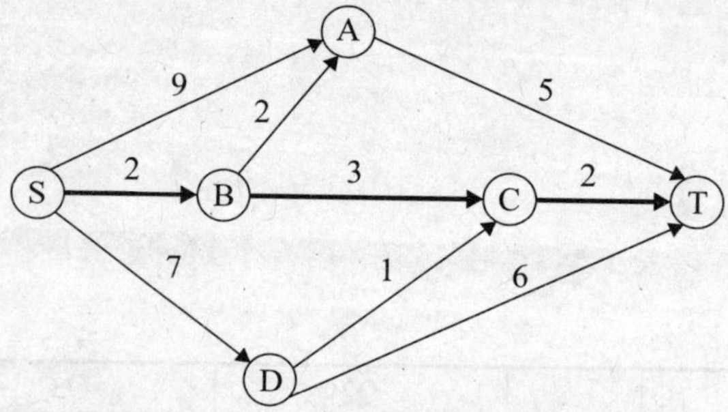

Motorists travelling from point S to point T would obviously take the route for which the total cost of traveling is the minimum. If two or more routes have the same least travel cost, then motorists are indifferent between them. Hence, the traffic gets evenly distributed among all the least cost routes.

The government can control the flow of traffic only by levying appropriate toll at each junction. For example, if a motorist takes the route S-A-T (using junction A alone), then the total cost of travel would be Rs 14 (i.e., Rs 9 + Rs 5) plus the toll charged at junction A.

Q1. If the government wants to ensure that no traffic flows on the street from D to T, while equal amount of traffic flows through junctions A and C, then a feasible set of toll charged (in rupees) at junctions A, B, C, and D respectively to achieve this goal is:

(a) 1, 5, 3, 3 (b) 1, 4, 4, 3 (c) 1, 5, 4, 2 (d) 0, 5, 2, 2

Q2. If the government wants to ensure that all motorists travelling from S to T pay the same amount (fuel costs and toll combined) regardless of the route they choose and the street from B to C is under repairs (and hence unusable), then a feasible set of toll charged (in rupees) at junctions A, B, C, and D respectively to achieve this goal is:

(a) 2, 5, 3, 2 (b) 1, 5, 3, 2 (c) 2, 3, 5, 1 (d) 1, 3, 5, 1

Q3. If the government wants to ensure that the traffic at S gets evenly distributed along streets from S to A, from S to B, and from S to D, then a feasible set of toll charged (in rupees) at junctions A, B, C, and D respectively to achieve this goal is:

(a) 0, 5, 4, 1 (b) 0, 5, 2, 2 (c) 1, 5, 3, 3 (d) 1, 5, 3, 2 (e) 0, 4, 3, 2

Q4. If the government wants to ensure that all routes from S to T get the same amount of traffic, then a feasible set of toll charged (in rupees) at junctions A, B, C, and D respectively to achieve this goal is:

(a) 0, 5, 2, 2 (b) 0, 5, 4, 1 (c) 1, 5, 3, 3 (d) 1, 5, 3, 2 (e) 1, 5, 4, 2

Q5. The government wants to devise a toll policy such that the total cost to the commuters per trip is minimized. The policy should also ensure that not more than 70 per cent of the total traffic passes through junction B. The cost incurred by the commuter travelling from point S to point T under this policy will be:

(a) Rs 7 (b) Rs 9 (c) Rs 10 (d) Rs 13 (e) Rs 14

Rough Work

Answer Grid

Q1	Q2	Q3	Q4	Q5
c	c	a	b	d

Solutions to Q1–Q5

A1. Assume that the toll tax at A, B, C and D be *a, b, c* and *d* respectively. For this question we are required to equalize the cost on each road except SDT

Hence, $14 + a = 9 + a + b = 7 + b + c = 10 + d + c$

Or, $b = 5$

Now, $14 + a = 12 + c = 10 + d + c$

Or, $d = 2$ and $c - a = 2$

Now, from the answer options only option (d) satisfies the above values so the correct option is (d).

Alternatively this question can be done by options also.

A2. Now for this question we have to equalize the cost on each road except SBCT

So, $14 + a = 9 + a + b = 10 + d + c = 13 + d$

Or, $b = 5$ and $c = 3$

Now, $14 + a = 13 + d$

Or, $d - a = 1$.

Option (b) satisfies the conditions.

A3. So as to minimize the cost the traffic will be equally divide between the roots SAT, SBAT and SDT. Hence the answer option is (a).

A4. Minimum cost considering toll tax on any of the given four junctions = Zero. We can say that 100% traffic will be attracted by the route S-B-C-T(Total cost – 7). If the cost of this route is increased to 8, condition will remain same. Again if we increase it to 9 then traffic will get distributed into two routes S-B-C-T and S-B-A-T as the costs are same. However still B caters 100% traffic. In order to avoid this, the respective costs of these two routes are increased to 10 then the traffic will get distributed into three routes S-B-C-T,S-B-A-T and S-D-C-T. It can be seen now that traffic will get distributed evenly means 33.33%in each route. Total traffic flow through B = 33.33% + 33.33% = 66.66% that is less than 70%. Hence answer is option (c).

A5. Answer is option (c).

Chetan and Michael

Set 67

Direction for questions 1–5

Read the information given below and solve the questions based on it.

Two traders, Chetan and Michael, were involved in the buying and selling of MCS shares over five trading days. At the beginning of the first day, the MCS share was priced at Rs 100, while at the end of the fifth day it was priced at Rs 110. At the end of each day, the MCS share price either went up by Rs 10, or else, it came down by Rs 10. Both Chetan and Michael took buying and selling decisions at the end of each trading day.

The beginning price of MCS share on a given day was the same as the ending price of the previous day.

Chetan and Michael started with the same number of shares and amount of cash, and had enough of both.

Below are some additional facts about how Chetan and Michael traded over the five trading days.

- Each day if the price went up, Chetan sold 10 shares of MCS at the closing price. On the other hand, each day if the price went down, he bought 10 shares at the closing price.
- If on any day, the closing price was above Rs 110, then Michael sold 10 shares of MCS, while if it was below Rs 90, he bought 10 shares, all at the closing price.

Q1. If Chetan sold 10 shares of MCS on three consecutive days, while Michael sold 10 shares only once during the five days, what was the price of MCS at the end of day 3?

(a) Rs 90 (b) Rs 100 (c) Rs 110 (d) Rs 120 (e) Rs 130

Q2. If Chetan ended up with Rs 1300 more cash than Michael at the end of day 5, what was the price of MCS share at the end of day 4?

(a) Rs 90 (b) Rs 100 (c) Rs 110 (d) Rs 120 (e) Not uniquely determinable

Q3. If Michael ended up with 20 more shares than Chetan at the end of day 5, what was the price of the share at the end of day 3?

(a) Rs 90 (b) Rs 100 (c) Rs 110 (d) Rs 120 (e) Rs 130

Q4. If Michael ended up with Rs 100 less cash than Chetan at the end of day 5, what was the difference in the number of shares possessed by Michael and Chetan (at the end of day 5)?

(a) Michael had 10 less shares than Chetan.
(b) Michael had 10 more shares than Chetan.
(c) Chetan had 10 more shares than Michael,
(d) Chetan had 20 more shares than Michael.
(e) Both had the same number of shares.

Q5. What could have been the maximum possible increase in combined cash balance of Chetan and Michael at the end of the fifth day?

(a)Rs 3700 (b) Rs 4000 (c) Rs 4700 (d) Rs 5000 (e) Rs 6000

Rough Work

Answer Grid

Q1	Q2	Q3	Q4	Q5
c	b	a	e	d

Solutions to Q1–Q5

On 1st day, price was Rs100 per share. Now we can have the following cases:

Cases	1	2	3	4	5	6	7	8	9	10
At the end of 1st day	90	110	90	110	110	90	110	110	90	110
At the end of 2nd day	100	100	80	120	100	100	120	120	100	120
At the end of 3rd day	90	90	90	110	110	110	130	110	110	110
At the end of 4th day	100	100	100	100	100	100	120	120	120	120
At the end of 5th day	110	110	110	110	110	110	110	110	110	110

A1. Obviously from case-1, at the end of 3rd day, price of share is Rs110. Hence, the option (c) is the answer.

A2. Obviously from case-5, we get price of share at the end of Day 4 is Rs100 Hence, the option (b) is the answer.

A3. Obviously from case-3, we get price of share at the end of day 3 is Rs90. Hence, the option (a) is the answer.

A4. Obviously from cases 5 and 9, we have

	Chetan	Michael
110	+10	0
120	+10	+10
110	−10	0
100	−10	0
110	+10	0

	Chetan	Michael
90	−10	0
100	+10	0
110	+10	0
120	+10	+10
110	−10	0

Using 1st table, Chetan has invested Rs1300 and Michael has invested Rs1200.
Michael has Rs100 less than chetan, so both have same no. of shares.
Hence, the option (e) is the answer.

A5. Obviously from case-7, we get maximum possible increase in combined cash balance at the end of 5th day, that is (1200 + 1300 + 1200) = 3700 for Chetan and (−1100 − 1200 + 1300 − 1100 − 1200) = 1300 for Michael
Hence maximum possible increase in combined balance = 3700 + 1300 = Rs5000.
Hence, the option (d) is the answer

Team Selection

Set 68

Direction for questions 1–5

K, L, M, N, P, Q, R, S, U and W are the only ten members in a department. There is a proposal to form a team from within the members of the department, subject to the following conditions:

- A team must include exactly one among P, R, and S.
- A team must include either M or Q, but not both.
- If a team includes K, then it must also include L, and vice versa.
- If a team includes one among S, U, and W, then it must also include the other two.
- L and N cannot be members of the same team.
- L and U cannot be members of the same team.
- The size of a team is defined as the number of members in the team.

Q1. Who cannot be a member of a team of size 3?

(a) L (b) M (c) N (d) P (e) Q

Q2. Who can be a member of a team of size 5?

(a) K (b) L (c) M (d) P (e) R

Q3. What would be the size of the largest possible team?

(a) 8 (b) 7 (c) 6 (d) 5 (e) Cannot be determined

Q4. What could be the size of a team that includes K?

(a) 2 or 3 (b) 2 or 4 (c) 3 or 4 (d) Only 2 (e) Only 4

Q5. In how many ways a team can be constituted so that the team includes N?

(a)2 (b) 3 (c) 4 (d) 5 (e) 6

Rough Work

Answer Grid

Q1	Q2	Q3	Q4	Q5
a	c	d	e	e

Solutions to Q1–Q5

A1. In the team of size 3, one member will be from P, R and S and one member will be from M and Q.
Now only one member has to be determined and we know that K and L will be always in team and since there is place for only 1 member, hence K and L will not be in team of 3 members.
Hence, the option (a) is the answer.

A2. If we take K and L we cannot take N and U so the size of team will be 4 only. Now we will take S from first statement so U and W will also be there. Now as U so there L cannot be there and hence K will also not be there. As L is not in team so we can take N so 4 members of team are S, U, W and N and because of second statement any one of M or Q will be there and only M is given in the options
Hence, the option (c) is the answer.

A3. Using first statement, if we take P or R, U and W cannot be in team. To get maximum members in team we should take S and from 4th statement U and W will also be in the team. Now either M or Q will be included in the team, and since U is there L can't be included in the team hence K will also not be there.
Now as L is not in the team, N can be included in the team. So team with maximum numbers of members will be S, W, U, N and any one from M and Q.
Hence maximum possible size of team of team is 5.
Hence, the option (d) is the answer.

A4. If we have to take K, L will always be there so U and N will not be in the team. As U is not in the team so S and W will also not be included in the team.
As anyone from P and R will always be there in team and same in the case with M and Q, so the size of team that include K will always be 4.
Hence, the option (e) is the answer.

A5. Following are the ways in which a team can be constituted:
(1) P,M,N
(2) P,Q,N
(3) R,M,N
(4) R,Q,N
(5) S,U,W,N,M
(6) S,U,W,N,Q
Hence the correct answer option is (e).

Management Institute Faculty Profile

Set 69

Direction for questions 1–4

A management institute was established on January 1, 2000 with 3, 4, 5, and 6 faculty members in the Marketing, Organisational Behaviour (OB), Finance and Operations Management (OM) areas respectively, to start with. No faculty member retired or joined the institute in the first three months of the year 2000. In the next four years, the institute recruited one faculty member in each of the four areas. All these new faculty members, who joined the institute subsequently over the years, were 25 years old at the time of their joining the institute. All of them joined the institute on April 1. During these four years, one of the faculty members retired at the age of 60. The following diagram gives the area-wise average age (in terms of number of completed years) of faculty members as on April 1 of 2000, 2001, 2002, and 2003.

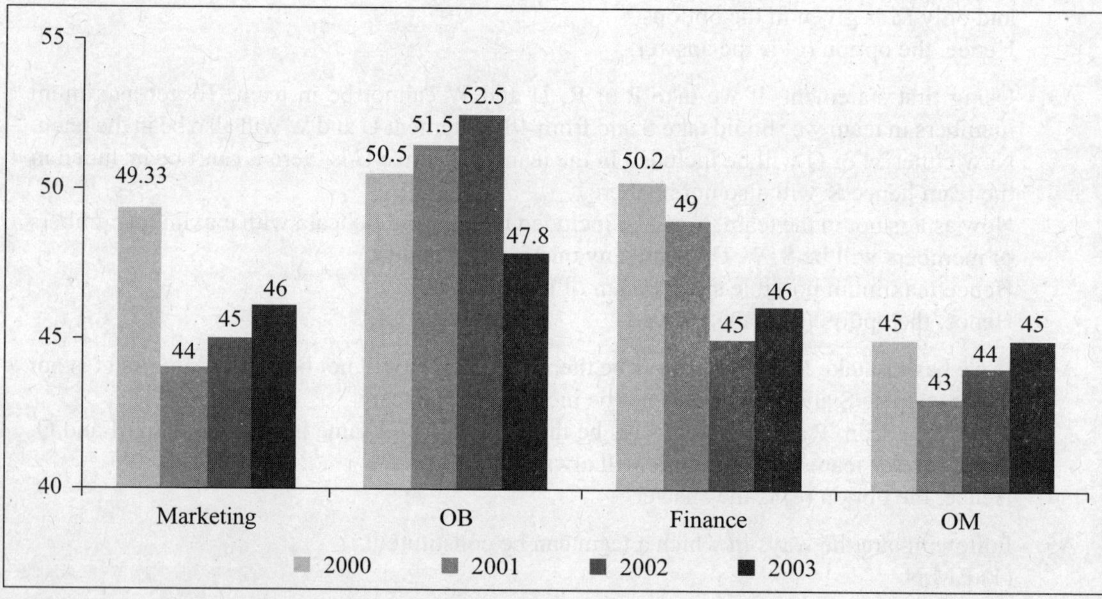

Q1. From which area did the faculty member retire?

(a) Finance (b) Marketing (c) OB (d) OM

Q2. Professors Naresh and Devesh, two faculty members in the Marketing area, who have been with the Institute since its inception, share a birthday, which falls on 20thNovember. One was born in 1947 and the other one in 1950. On April 1 2005, what was the age of the third faculty member, who has been in the same area since inception?

(a) 47 (b) 50 (c) 51 (d) 52

Q3. In which year did the new faculty member join the Finance area?

(a) 2000 (b) 2001 (c) 2002 (d) 2003

Q4. What was the age of the new faculty member, who joined the OM area, as on April 1, 2003?

(a) 25 (b) 26 (c) 27 (d) 28

Rough Work

Answer Grid

Q1	Q2	Q3	Q4
a	d	c	d

Solutions to Q1–Q3

A1. There are two dips in the Finance only. Hence finance is the area where the Prof. Retire from. Hence option (a) is the answer.

A2. Age of Prof. Naresh + Age of Prof. Devesh = 101
Sum of ages of all the professors of marketing at the time of inception = 148
Hence the age of 3rd professor at the time of inception = 47
Hence his age in 2005 = 52 years. So, option (d) is the answer.

A3. In 2001, one professor retires from the Finance area and the next year 2002, a new professor joins Finance area. Hence option (c) is the answer.

Disguised Identities

Set 70

Direction for questions 1–4

Go through the data set given below and solve the questions based on it.

The table below presents the revenue (in million rupees) of four firms in three states. These firms, Honest Ltd, Aggressive Ltd, Truthful Ltd and Profitable Ltd are disguised in the table as A, B, C and D, in no particular order.

States	Firm A	Firm B	Firm C	Firm D
UP	49	82	80	55
Bihar	69	72	70	65
MP	72	63	72	65

Further, it is known that:

- In the state of MP, Truthful Ltd has the highest market share.
- Aggressive Ltd's aggregate revenue differs from Honest Ltd's by Rs 5 million.

Q1. What can be said regarding the following two statements?

Statement 1: Profitable Ltd has the lowest share in MP market.

Statement 2: Honest Ltd's total revenue is more than Profitable Ltd

(a) If Statement 1 is true then Statement 2 is necessarily true.

(b) If Statement 1 is true then Statement 2 is necessarily false.

(c) Both Statement 1 and Statement 2 are true.

(d) Neither Statement 1 nor Statement 2 is true.

Q2. What can be said regarding the following two statements?

Statement 1: Aggressive Ltd's lowest revenues are from MP.

Statement 2: Honest Ltd's lowest revenues are from Bihar.

(a) If Statement 2 is true then Statement 1 is necessarily false.

(b) If Statement 1 is false then Statement 2 is necessarily true.

(c) If Statement 1 is true then Statement 2 is necessarily true.

(d) None of the above.

Q3. What can be said regarding the following two statements?

Statement 1: Honest Ltd has the highest share in the UP market.

Statement 2: Aggressive Ltd has the highest share in the Bihar market.

(a) Both statements could be true.

(b) At least one of the statements must be true.

(c) At most one of the statements is true.

(d) None of the above.

Q4. If Profitable Ltd's lowest revenue is from UP, then which of the following is true?

(a) Truthful Ltd's lowest revenues are from MP.

(b) Truthful Ltd's lowest revenues are from Bihar.

(c) Truthful Ltd's lowest revenues are from UP.

(d) No definite conclusion is possible.

Rough Work

Answer Grid

Q1	Q2	Q3	Q4
b	c	c	c

Solutions to Q1–Q4

States	Firm A	Firm B	Firm C	Firm D
UP	49	82	80	55
Bihar	69	72	70	65
MP	72	63	72	65
Total	190	217	222	185

Combination of Aggressive Ltd and Honest Ltd is possible only for :- Firm A + Firm D or Firm B and Firm C.
And Truthful Ltd can be only Firm A or Firm C.

A1. According to statement 1, B is profitable, then A and D are aggressive and honest. Then honest total revenue cannot be more than that of profitable. So statement 2 is not true. Hence option (b) is the answer.

A2. According to statement 1 aggressive is Firm (B). So, honest Ltd will be C. Then statement 2 is also true. So, Honest Ltd's. Lowest revenue is from Bihar.
Hence option (c) is the answer.

A3. Firm B is Honest Ltd (according to statement 1), then statement 2 is false. At the same time, it also can be seen that both the statements can be simultaneously wrong. Hence "At most" one of the statements is true. Hence option (c) is the answer.

A4. Profitable can be either A or D. Then aggressive and honest has to be B and C. hence truthful Ltd is either Firm A or D. And for both A and D lowest revenue is from UP. Hence option (c) is the answer.

Venkat's Income

Set 71

Direction for questions 1–4

Venkat, a stockbroker, invested a part of his money in the stock of four companies—A, B, C and D. Each of these companies belonged to different industries, viz., Cement, Information Technology (IT), Auto and Steel, in no particular order. At the time of investment, the price of each stock was Rs 100. Venkat purchased only one stock of each of these companies. He was expecting returns of 20%, 10%, 30%, and 40% from the stock of companies A, B, C and D, respectively. Returns are defined as the change in the value of the stock after one year, expressed as a percentage of the initial value. During the year, two of these companies announced extraordinarily good results. One of these two companies belonged to the Cement or the IT industry, while the other one belonged to either the Steel or the Auto industry. As a result, the returns on the stocks of these two companies were higher than the initially expected returns. For the company belonging to the Cement or the IT industry with extraordinarily good results, the returns were twice that of the initially expected returns. For the company belonging to the Steel or the Auto industry, the returns on announcement of extraordinarily good results were only one and a half times that of the initially expected returns. For the remaining two companies, which did not announce extraordinarily good results, the returns realized during the year were the same as initially expected.

Q1. What is the minimum average return Venkat would have earned during the year?

(a) 30% (b) 31 ¼ % (c) 32 ½ % (d) Cannot be determined

Q2. If Venkat earned a 35% return on average during the year, then which of these statements would necessarily be true?

I. Company A belonged either to Auto or to Steel industry.
II. Company B did not announce extraordinarily good results.
III. Company A announced extraordinarily good results.
IV. Company D did not announce extraordinarily good results.

(a) I and II only (b) II and III only (c) III and IV only (d) II and IV only

Q3. If Venkat earned a 38.75% return on average during the year, then which of these statement(s) would necessarily be true?

I. Company C belonged either to Auto or to Steel industry.
II. Company D belonged either to Auto or to Steel industry.
III. Company A announced extraordinarily good results.
IV. Company B did not announce extraordinarily good results.

(a) I and II only (b) II and III only (c) I and IV only (d) II and IV only

Q4. If Company C belonged to the Cement or the IT industry and did announce extraordinarily good results, then which of these statement(s) would necessarily be true?

I. Venkat earned not more than 36.25% return on average.
II. Venkat earned not less than 33.75% return on average.
III. If Venkat earned 33.75% return on average, Company A announced extraordinarily good results.
IV. If Venkat earned 33.75% return on average, Company B belonged either to Auto or to Steel industry.

(a) I and II only (b) II and IV only (c) II and III only (d) III and IV only

Rough Work

Answer Grid

Q1	Q2	Q3	Q4
a	b	c	b

Solutions to Q1–Q4

A1. We will have the minimum returns if the extraordinary performing stocks (double and 1.5 growth) are the one's whose expected returns are lowest (i.e., 10% and 20%). Taking the minimum value of the expected returns as 10%, we will be required to see that which of the two values of 10 and 20 multiplied by 2 and 1.5 and vice-versa will give the minimum value.

Hence comparing the minimum value between $20 \times 2 + 10 \times 1.5$ and $20 \times 1.5 + 10 \times 2$, we find that the 2nd one is minimum.

So, the minimum average return is $\frac{20 \times 1.5 + 10 \times 2 + 30 + 40}{4} = 30\%$

Hence option (a) is the answer.

A2. Total return is $35 \times 4 = 140$ (Average return is 35%)

There is only one possible arrangement of 140 ($40 \times 1.5 + 30 + 20 \times 2 + 10$)

A = 20×2 (Cement or IT)

B = 10

C = 30

D = 40(1.5) (Steel or Auto)

Using the data given in the question, we can see that A has to be cement or IT, D is steel or Auto.

Hence statement (II) and (III) are correct.

A3. Total return is 38.75 × 4 = 155. This is the maximum possible return that Venkat can have.
The possible arrangement is 20 + 10 + 30 × 1.5 + 40 × 2
Hence A = 20, B = 10, C= 30(steel or Auto)
D = 40 (cement or IT)
So, Statement (i) and (iv) are correct.
So, option (c) is the answer.

A4. Give C. ……. Cement or IT industry
C's return is 30 × 2 = 60%
Among the other values we see that following are the possible arrangements:
10 × 1.5 + 20 + 40, 10 + 20 × 1.5 + 40, 40 + 20 + 40 × 1.5
The average returns will be in each case

$$\frac{10 \times 1.5 + 20 + 40 + 60}{4} \quad (33.75\%)$$

$$\frac{10 + 20 \times 1.5 + 40 + 60}{4} \quad (35\%)$$

Considering 33.75% as the valid value, then B belongs to the auto industry.
So, (II) and (IV) are correct.
So, option (b) is the correct answer.

Tennis Tournament Set 72

Direction for questions 1–4

In the table below is the listing of players, seeded from highest (#1) to lowest (#32), who are due to play in an Association of Tennis Players (ATP) tournament for women. This tournament has four knockout rounds before the final, i.e., first round, second round, quarterfinals, and semi-finals. In the first round, the highest seeded player plays the lowest seeded player (seed # 32) which is designated match No. 1 of first round; the 2nd seeded player plays the 31st seeded player which is designated match No. 2 of the first round, and so on. Thus, for instance, match No. 16 of first round is to be played between 16th seeded player and the 17th seeded player. In the second round, the winner of match No. 1 of first round plays the winner of match No. 16 of first round and is designated match No. 1 of second round. Similarly, the winner of match No. 2 of first round plays the winner of match No. 15 of first round, and is designated match No. 2 of second round. Thus, for instance, match No. 8 of the second round is to be played between the winner of match No. 8 of first round and the winner of match No. 9 of first round. The same pattern is followed for later rounds as well.

Seed #	Name of Player	Seed #	Name of Player	Seed #	Name of Player
1	Maria Sharapova	12	Mary Pierce	23	Silvia Farina Elia
2	Lindsay Davenport	13	Anastasia Myskina	24	Tatiana Golovin
3	Amelie Mauresmo	14	Alicia Molik	25	Shinobu Asagoe
4	Kim Clijsters	15	Nathalie Dechy	26	Francesca Schiavonc
5	Svetlana Kuznetsova	16	Elena Bovina	27	Nicole Vaidisova
6	Elena Dementieva	17	Jelena Jankovic	28	Gisela Dulko
7	Justine Henin	18	Ana Ivanovic	29	Flavia Pennetta
8	Serena Williams	19	Vera Zvonareva	30	Anna Chakvetadze
9	Nadia Petrova	20	Elena Likhovtseva	31	Ai Sugiyama
10	Venus Williams	21	Daniela Hantuchova	32	Anna-lena Groenefeld
11	Patty Schnyder	22	Dinara Safina		

Q1. If there are no upsets (a lower seeded player beating a higher seeded player) in the first round, and only match Nos. 6, 7, and 8 of the second round result in upsets, then who would meet Lindsay Davenport in quarter finals, in case Davenport reaches quarter finals?
(a) Justine Henin (b) Nadia Petrova (c) Patty Schnyder
(d) Venus Williams

Q2. If Elena Dementieva and Serena Williams lose in the second round, while Justine Henin and Nadia Petrova make it to the semi-finals, then who would play Maria Sharapova in the quarter finals, in the event Sharapova reaches quarter finals?
(a) Dinara Safina (b) Justine Henin (c) Nadia Petrova
(d) Patty Schnyder

Q3. If, in the first round, all even numbered matches (and none of the odd numbered ones) result in upsets, and there are no upsets in the second round, then who could be the lowest seeded player facing Maria Sharapova in semi-finals?

(a) Anastasia Myskina (b) Flavia Pennetta (c) Nadia Petrova
(d) Svetlana Kuznetsova

Q4. If the top eight seeds make it to the quarter finals, then who, amongst the players listed below, would definitely not play against Maria Sharapova in the final, in case Sharapova reaches the final?

(a) Amelie Mauresmo (b) Elena Dementieva (c) Kim Clijsters
(d) Lindsay Davenport

Answer Grid

Q1	Q2	Q3	Q4
c	b	b	a

Solutions to Q1–Q4

A1. We will be required to draw two tables. Winners after round two would be 1, 2, 3, 4, 5, 11, 10, 9 for 8 rounds respectively. Since Lindsay is number two she will play Venus Williams in quarter-final.
Hence option (d) is the answer.

A2. Elena is at number 6, Serena Williams is at number 8.
Maria is number 1 she will play number 9, i.e., Nadia Petrova
Hence option (c) is the answer.

A3.

1	32
2	31
3	30
4	29
5	28
6	27
7	26
8	25
9	24
10	23
11	22
12	21
13	20
14	19
15	18
16	17

Then from the table, would be winners are: 1, 31, 29, 5, 27, 7, 25, 9, 23, 11, 21, 12, 19, 15, 17. So the table for the next round would be:

1	17
31	15
3	19
29	13
5	21
27	11
7	23
25	9

Since there is no upset in second round, hence table in next round will be like:

1	9
15	7
3	11
13	5

We have Maria Sharapova in the semi-final. Since we are not sure what is the result of other games, table is drawn as given below:
Table in next round would be like:

1	5/13
7/15	3/11

Hence option (a) is the answer.

A4.

1	8
2	7
3	6
4	5

In this case, Kim Clijster will either not reach semi-final or she will play Maria Sharapova in semi-final.
In that case, one of these two will be eliminated. Hence she cannot play Maria Sharapova in the finals, as Maria Sharapova is playing in final.
Hence option (c) is the answer.

Prof. Singh's Homepage Set 73

Direction for questions 1–4

Answer the questions on the basis of the information given below.

Prof. Singh has been tracking the number of visitors to his homepage. His service provider has provided him with the following data on the country of origin of the visitors and the university they belong to:

Number of visitors

	DAY		
Country	1	2	3
Canada	2	0	0
Netherlands	1	1	0
India	1	2	0
UK	2	0	2
USA	1	0	1

Number of visitors

	DAY		
UNIVERSITY	1	2	3
University 1	1	0	0
University 2	2	0	0
University 3	0	1	0
University 4	0	0	2
University 5	1	0	0
University 6	1	0	1
University 7	2	0	0
University 8	0	2	0

Q1. To which country does university 5 belong?
(a) India or Netherlands but not USA (b) India or USA but not Netherlands
(c) Netherlands or USA but not India (d) India or USA but not UK.

Q2. University 1 can belong to
(a) UK (b) Canada (c) Netherlands (d) USA

Q3. Which among the listed countries can possibly host three of the eight listed universities?
(a) None (b) Only UK (c) Only India (d) Both India and UK

Q4. Visitors from how many universities from UK visited Prof. Singh's homepage in the three days?
(a) 1 (b) 2 (c) 3 (d) 4

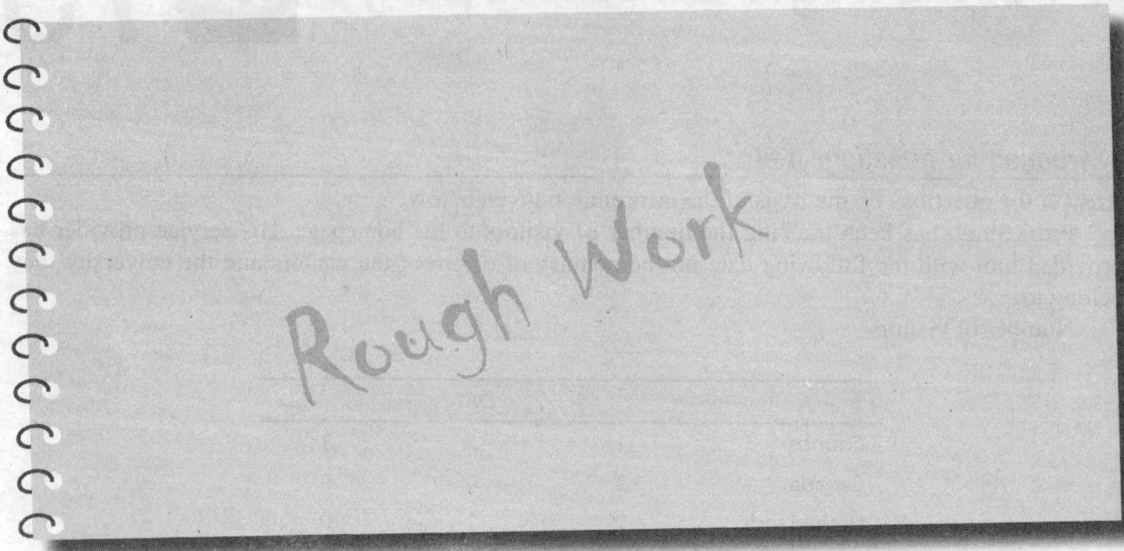

Answer Grid

Q1	Q2	Q3	Q4
a	c	a	b

Solutions to Q1–Q4

When we compare table 1 and 2, university 4 corresponds to UK and university 6 corresponds to USA (after day 3 onwards values are concerned and university 8 corresponds to India and university 3 to Netherlands and now Indian or Netherlands can take university 1 or university 5. Now university 2 and 7 belongs to either UK or Canada (Only one).

University	Day			Country
	1	2	3	
University 1	1	0	0	India/Netherlands
University 2	2	0	0	UK/Canada
University 3	0	1	0	Netherlands
University 4	0	0	2	UK
University 5	1	0	0	India/Netherlands
University 6	1	0	1	USA
University 7	2	0	0	UK/Canada
University 8	0	2	0	India

Now all the questions can be answered.

A1. University 5 can belong to either India or Netherlands, but not USA. Hence option (a) is the answer.

A2. University 1 can belong either to Netherlands or India. Hence option (c) is the answer.

A3. None of the countries can host three universities. Hence option (a) is the answer.

A4. University 4 and either University 2 or 7 belong to UK. Hence there are two universities that belong to UK. Hence option (b) is the answer.

Batting Parameters

Set 74

Direction for questions 1–4

Answer the questions on the basis of the information given below.

Coach John sat with the score cards of Indian players from the 3 games in a one day cricket tournament where the same set of players played for India and all the major batsman got out. John summarized the batting performance through three diagrams, one for each game. In each diagram, the three outer triangles communicate the number of runs scored that was scored by the three top scores from India, where K,R,S,V, and Y represents Kaif, Rahul, Saurav, Virender and Yuvraj respectively. The middle triangle in each diagram denotes the percentage of the total score that was scored by the top three Indian scorers in that game. No two players score the same number of runs in the same game. John also calculated two batting indices for each player based on his scores in the tournaments; the R-index of a batsman is the difference between his highest and lowest scores in the 3 games while the M-index is the middle number, if his scores are arranged in a non-increasing order.

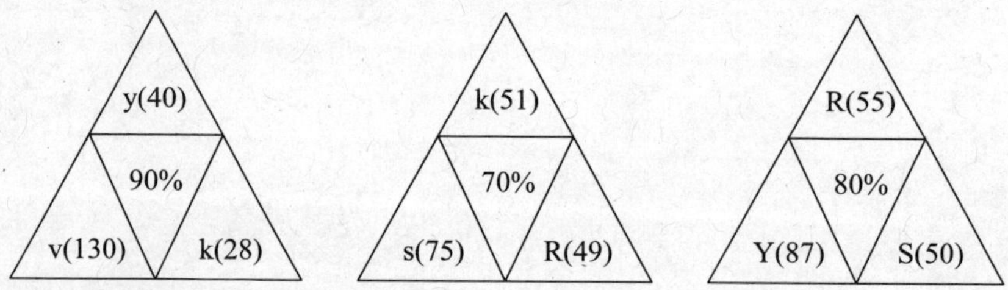

Q1. For how many Indian players is it possible to calculate the exact M-index?
(a) 0 (b) 1 (c) 2 (d) More than 2

Q2. Among the players mentioned , who can have the lowest R-index from the tournament?
(a) Only Kaif, Rahul or Yuvraj (b) Only Kaif or Rahul
(c) Only Kaif or Yuvraj (c) Only Kaif

Q3. How many players among those listed definitely scored less than Yuvraj in the tournament?
(a) 0 (b) 1 (c) 2 (d) More than 2

Q4. Which of the players had the best M-index from the tournament?
(a) Rahul (b) Saurav (c) Virender (d) Yuvraj

Rough Work

Answer Grid

Q1	Q2	Q3	Q4
c	a	b	b

Solutions to Q1–Q4

Let us have a tabular presentation of the data given:

	Match 1	Match 2	Match 3
K	28	51	<48
R	<22	49	55
S	<22	75	50
V	130	<49	<48
Y	40	<49	87
Top 3 batsman	198	175	192
India total	220	250	240

Now all the questions can be answered.

A1. (c)

A2. (a)

A3. (b)

A4. (b)

UN Conference

Set 75

Direction for questions 1–4

Answer the question on the basis of the information given below.

Twenty one participants from four continents (Africa, America, Australasia, and Europe) attended a United Nations conference. Each participant was an expert in one of the four fields—labour, health, population studies and refugee relocation. The following five facts about the participants are given.

(a) The number of labour experts in the camp was exactly the half the number of experts in each of the other three fields.

(b) Africa did not send any labour expert. Otherwise every continent, including Africa, sent at least one expert in each category.

(c) None of the continents sent more than three experts in any category.

(d) If there had been one less Australian expert, then the Americans would have had twice as many experts as each of the other continents.

(e) Mike and Alfonzo are leading experts of population studies who attended the conference. They are from Australasia.

Q1. Which of the following combination is NOT possible?

(a) 2 experts in population studies from America and 2 health experts from Africa attended the conference.

(b) 2 experts in population studies from America and 1 health expert from Africa attended the conference.

(c) 3 experts in refugee relocation from America and 1 health expert from Africa attended the conference.

(d) Africa and America each had 1 expert in population studies attending the conference.

Q2. If Ramos is the lone American expert in population studies, which of the following is NOT true about the numbers of experts in the conference from the four continents?

(a) There is one expert in health from America.

(b) There is one expert in refugee relocation from Africa.

(c) There are two experts in health from America.

(d) There are three experts in refugee relocation from America.

Q3. Alex, an American expert in refugee relocation, was the first keynote speaker in the conference. What can be inferred the number about the number of American experts in refugee relocation in the conference, excluding Alex?

i. At least one

ii. At most two

(a) Only i and not ii
(b) only ii and not i
(c) Both i and not ii
(d) Neither i nor ii

Q4. Which of the following numbers cannot be determined from the information given?

(a) Number of labour experts from America.

(b) Number of health experts from Europe.

(c) Number of health experts from Australasia.

(d) Number of experts in refugee relocation from Africa.

Rough Work

Answer Grid

Q1	Q2	Q3	Q4
d	c	c	d

Solutions to Q1–Q4

Let us have a tabular presentation of the data given:

	Africa	America	Australasia	Europe	
L	0	1	1	1	3
H			1	1	6
P			2	1	6
R			1	1	6
	4	8	5	4	
					21

(i) Because the number of labour experts is half of each of the others, the only possible combination is as follows:

L -3

H

P 6 each

R

(ii)

	Africa	America	Australasia	Europe	
L	0	1	1	1	3
H	2	2	1	1	6
P	1	2	2	1	6
R	1	3	1	1	6
	4	8	5	4	
					21

	Africa	America	Australasia	Europe	
L	0	1	1	1	3
H	1	3	1	1	6
P	1	2	2	1	6
R	2	2	1	1	6
	4	8	5	4	
					21

	Africa	America	Australasia	Europe	
L	0	1	1	1	3
H	1	3	1	1	6
P	2	1	2	1	6
R	1	3	1	1	6
	4	8	5	4	
					21

A1. Option (d) is the answer

A2. Option (c) is the answer

A3. Option (c) is the answer

A4. Option (d) is the answer

Unsolved Exercises

Number of Doctors – By Specialities

Set 76

Direction for questions 1–4

Go through the data sets given below and solve the questions based on it.

Following pie chart provides the number of Doctors in 2007:

Per cent of Doctors by Category

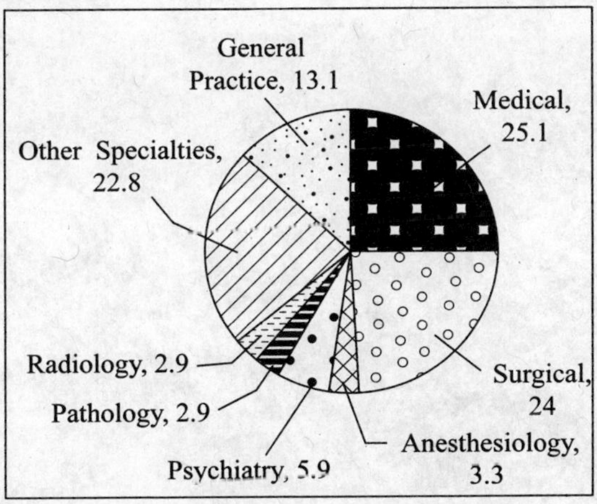

(100% = 4, 21, 300)

Number of Doctors within Medical and Surgical Categories (in thousands):

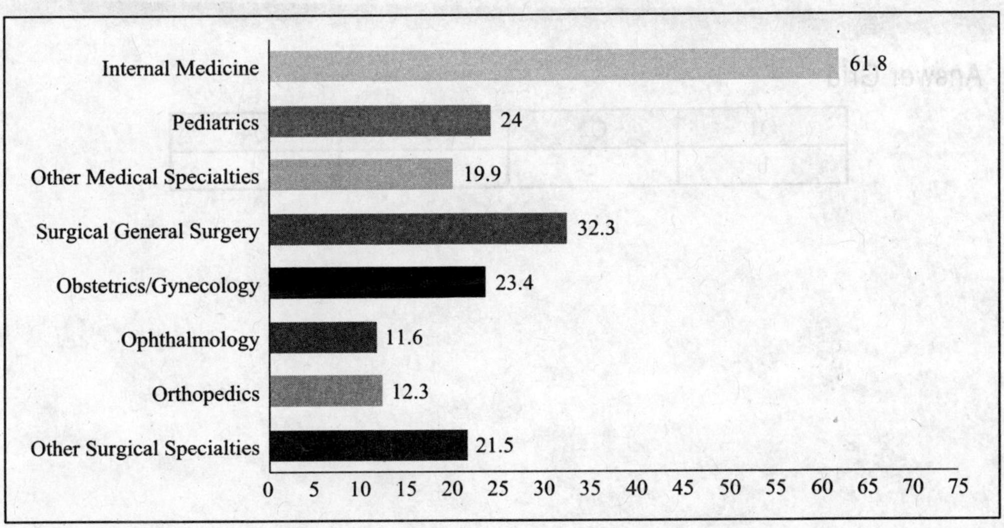

Internal Medicine, Pediatrics and Other Medical Specialties are Doctors within Medical categories. Rest all others are in Surgical Category.

Q1. Approximately what was the ratio of doctors in the surgical category to doctors in pathology?
(a) 10 : 1 (b) 8 : 1 (c) 7 : 1 (d) 5 : 6

Q2. Approximately how many more doctors were in psychiatry than in radiology?
(a) 3,000 (b) 6,300 (c) 12,600 (d) 24,800

Q3. Approximately how many of the doctors in the medical category were NOT in pediatrics?
(a) 61,800 (b) 76,000 (c) 81,700 (d) 92,600

Q4. If there was a total of 334,000 doctors in 2000, what was the approximate per cent increase in the number of doctors from 2000 to 2007?
(a) 10% (b) 12% (c) 16% (d) 26%

Rough Work

Answer Grid

Q1	Q2	Q3	Q4
b	c	c	d

Indigenous Population Set 77

Direction for questions 1–4

Go through the data set given below and solve the questions based on it.

The table gives data about estimated population of indigenous people selected countries in the year 1992.

Country	Indigenous population (in millions)	Share of national population (%)
Papua New Guinea	3	77
Bolivia	5.6	70
Guatemala	4.6	47
Peru	9	40
Ecuador	3.8	38
Myanmar	14	33
Laos	1.3	30
Mexico	10.9	12
New Zealand	0.4	12
Chile	0.2	9
Philippines	6	9
India	63	7
Malaysia	0.8	4
Canada	0.9	4
Australia	0.4	2
Brazil	1.5	1
Bangladesh	1.2	1
Thailand	0.5	1
United States	2	1
Former Soviet Union	1.4	<1

Q1. What is the population of Bolivia in the year 1994, if the population grows at an average of 4% p.a. in the country?

(a) 80 mn (b) 8 mn (c) 10.85 mn (d) 8.65 mn

Q2. If the indigenous population of Peru were to double in 1993 while the rest of population stagnates, what % of the total population of Peru will be of indigenous in nature?

(a) 57% (b) 80% (c) 66.67% (d) 72%

Q3. In 1992, what is the ratio of the total population of Canada to that of the United States?

(a) 6: 1 (b) 1: 9 (c) 1: 12 (d) 7.5: 1

Q4. For how many countries with indigenous population more than 5 million is the share of national population more than 33%?

(a) 1 (b) 2 (c) 3 (d) 4

Rough Work

Answer Grid

Q1	Q2	Q3	Q4
d	a	b	b

Decision Tree Analysis Set 78

Direction for questions 1–3

Rajat is a sales manager of Dubin Computers Ltd and looks after Delhi market. The company sells laptops in India. He is currently trying to select a distributor for coming five years. The distributor ensures that the products are accessible to the customers in the market. Market share of a company depends on the coverage by the distributor. The total profit potential of the entire laptop market in Delhi is Rs 5 crores in the current year and present value of next four years' cumulative profit potential is Rs 15 crores.

The first choice for Rajat is to enter into a long-term contract with a distributor M/s Jagan with whom Dubin has done business in the past, and whose distribution system reaches 55 per cent of all potential customers. At the last moment, however, a colleague suggests Rajat to consider signing a one-year contract with other distributors.

Distributors M/s Bola and M/s James are willing to be partners with Dubin. Although a year ago, M/s Bola's and M/s James's coverage reached only 40 and 25 per cent of customers respectively, they claim to have invested heavily in distribution resources and now expect to be able to reach 60 per cent and 75 per cent of customers respectively.

The probability of M/s Bola's claim and M/s James's claim to be true is 0.60 and 0.20 respectively. The knowledge about distributors' coverage will evolve over time. The assumption is that the true level of coverage offered by the new distributors could be discovered, with certainty, through a one-year trial, and this trial will reveal exactly one of the two levels of coverage: for example in case of M/s Bola—40 per cent (as it was last year) or 60 per cent (as claimed).

In addition, it is also assumed that whatever the coverage is for both distributors, it will not change over time. Rajat narrows down on three choices, which are as follows:

Choice 1. Give a five year contract to the familiar distributor M/s Jagan

Choice 2. Give a one year contract to the new distributor M/s Bola, and base next years, decision to renew contract with M/s Bola on observed coverage for next four years, contract with M/s Jagan.

Choice 3. Give a one-year contract to the new distributor M/s James, and base next year's decision to renew contract with M/s James on observed coverage for next four years, contract with M/s Jagan.

Q1. The expected present value of the five years cumulative profit with choice 3 is

(a) Rs 12.7 crores (b) Rs 10.6 crores
(c) Rs 11.7 crores (d) None of the above

Q2. Which of the following statement is TRUE?

(a) Choice 1 is more profitable than Choice 2
(b) Choice 3 is more profitable than Choice 2
(c) Choice 3 is more profitable than Choice 1
(d) None of the above

Q3. If the distributor M/s James claims coverage of 55% instead of 75% and probability of this claim to be true is 0.70 instead of 0.20, then which of the following statement is true?

(a) Choice 1 is more profitable than Choice 2
(b) Choice 2 is more profitable than Choice 3
(c) Choice 3 is more profitable than Choice 1
(d) None of the above

Rough Work

Answer Grid

Q1	Q2	Q3
b	d	b

Newspaper in Circulation Set 79

Direction for questions 1–5

Go through the data set given below and solve the questions based on it.

Following charts provide the details regarding the circulation of the newspaper *Times of Today*, its total yearly advertising revenue and average number of pages

Average daily circulation for newspaper *Times of Today*

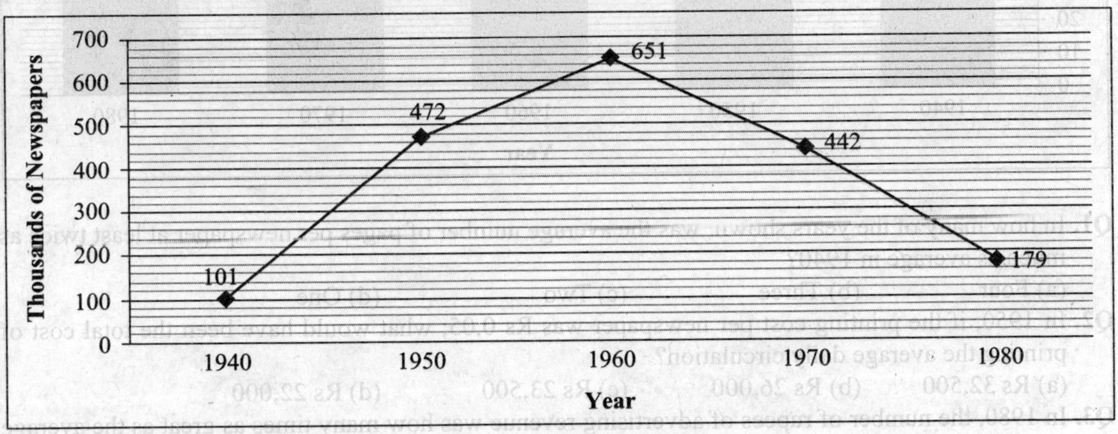

Total yearly advertising revenue for newspaper *Times of Today*

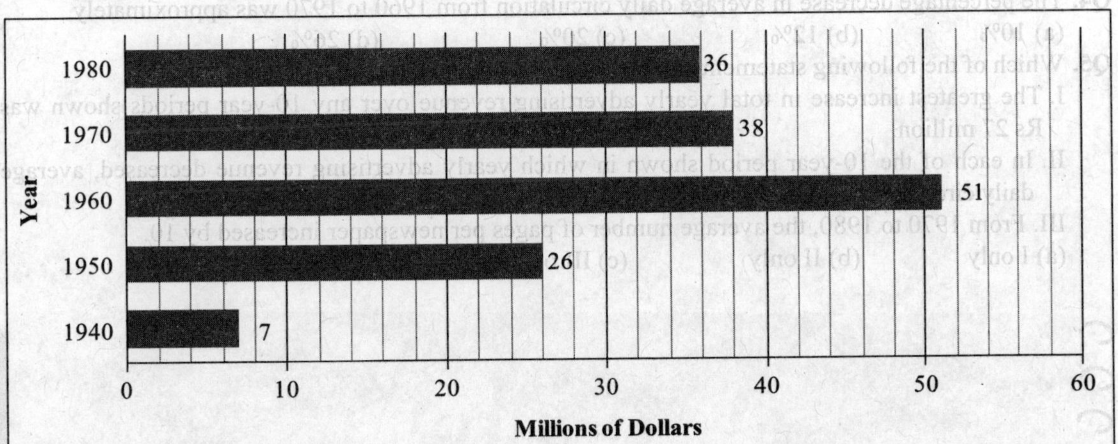

Average number of pages per copy for *Times of Today*

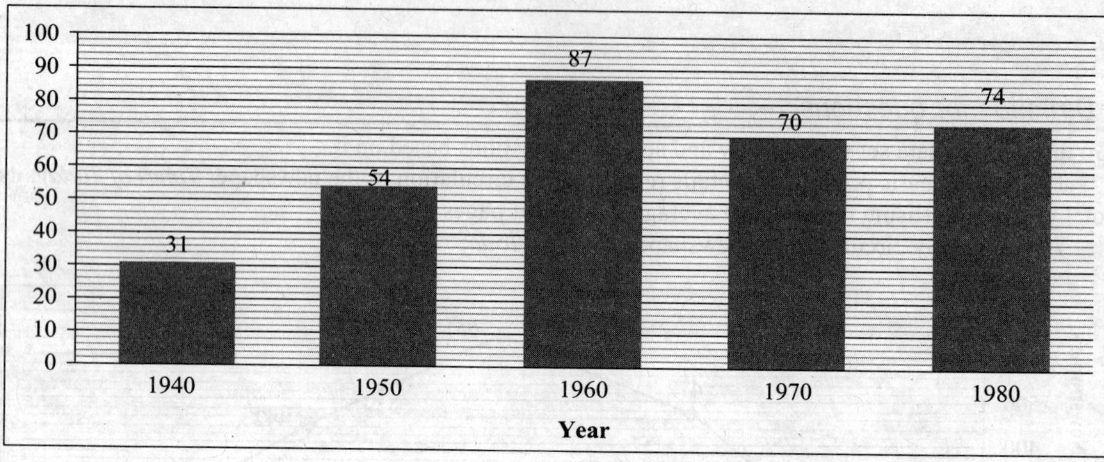

Q1. In how many of the years shown, was the average number of pages per newspaper at least twice as much as average in 1940?

(a) Four (b) Three (c) Two (d) One

Q2. In 1950, if the printing cost per newspaper was Rs 0.05, what would have been the total cost of printing the average daily circulation?

(a) Rs 32,500 (b) Rs 26,000 (c) Rs 23,500 (d) Rs 22,000

Q3. In 1980, the number of rupees of advertising revenue was how many times as great as the average daily circulation?

(a) 500 (b) 200 (c) 100 (d) 50

Q4. The percentage decrease in average daily circulation from 1960 to 1970 was approximately

(a) 10% (b) 12% (c) 20% (d) 26%

Q5. Which of the following statement can be inferred from the data?

I. The greatest increase in total yearly advertising revenue over any 10-year periods shown was Rs 27 million.

II. In each of the 10-year period shown in which yearly advertising revenue decreased, average daily circulation also decreased.

III. From 1970 to 1980, the average number of pages per newspaper increased by 10.

(a) I only (b) II only (c) III only (d) I and II

Rough Work

Answer Grid

Q1	Q2	Q3	Q4	Q5
b	c	b	d	d

Eligibility for the Workshops

Set 80

Direction for questions 1–3

Read the information given below and solve the questions based on it.

The table below reports the gender, designation and age group of the employees in an organization. It also provides information on their commitment to projects coming up in the months of January (Jan), February (Feb), March (Mar) and April (Apr), as well as their interest in attending workshops on: Business Opportunities (BO), Communication Skills (CS), and E-Governance (EG).

Sl. No.	Name	Gender	Designation	Age group	Committed to projects during	Interested in workshop on
1	Anshul	M	Mgr	Y	Jan, Mar	CS, EG
2	Bushkant	M	Dir	I	Feb, Mar	BO, EG
3	Charu	F	Mgr	I	Jan, Feb	BO, CS
4	Dinesh	M	Exe	O	Jan, Apr	BO, CS, EG
5	Eashwaran	M	Dir	O	Feb, Apr	BO
6	Fatima	F	Mgr	Y	Jan, Mar	BO, CS
7	Gayatri	F	Exe	Y	Feb, Mar	EG
8	Hari	M	Mgr	I	Feb, Mar	BO, CS, EG
9	Indira	F	Dir	O	Feb, Apr	BO, EG
10	John	M	Dir	Y	Jan, Mar	BO
11	Kalindi	F	Exe	I	Jan, Apr	BO, CS, EG
12	Lavanya	F	Mgr	O	Feb, Apr	CS, EG
13	Mandeep	M	Mgr	O	Mar, Apr	BO, EG
14	Nandlal	M	Dir	I	Jan, Feb	BO, EG
15	Parul	F	Exe	Y	Feb, Apr	CS, EG
16	Rahul	M	Mgr	Y	Mar, Apr	CS, EG
17	Sunita	F	Dir	Y	Jan, Feb	BO, EG
18	Urvashi	F	Exe	I	Feb, Mar	EG
19	Yamini	F	Mgr	O	Mar, Apr	CS, EG
20	Zeena	F	Exe	Y	Jan, Mar	BO, CS, EG

M=Male, F= Female; Exe=Executive, Mgr=Manager, Dir=Director; Y=Young, I=In-between, O=Old.

For each workshop, exactly four employees are to be sent, of which at least two should be females and at least one should be young. No employee can be sent to a workshop in which he/she is not interested in. An employee cannot attend the workshop on

- Communication Skills, if he/she is committed to internal projects in the month of January;
- Business Opportunities, if he/she is committed to internal projects in the month of February;
- E-governance, if he/she is committed to internal projects in the month of March.

Q1. Assuming that Parul and Hari are attending the workshop on Communication Skills (CS), then which of the following employees can possibly attend the CS workshop?

(a) Rahul and Yamini (b) Dinesh and Lavanya
(c) Anshul and Yamini (d) Fatima and Zeena

Q2. How many Executives (Exe) cannot attend more than one workshop?

(a) 2 (b) 3 (c) 15 (d) 16

Q3. Which set of employees cannot attend any of the workshops?

(a) Anshul, Charu, Eashwaran and Lavanya
(b) Anshul, Bushkant, Gayatri and Urvashi
(c) Charu, Urvashi, Bushkant and Mandeep
(d) Anshul, Gayatri, Eashwaran and Mandeep

Answer Grid

Q1	Q2	Q3
a	b	b

Car Sales

Set 81

Direction for questions 1–5

Go through the data set given below and solve the questions based on it.

Following bar chart and pie charts provide the data related to the total sales volume (in lacs) of new cars in India for the years 1990–2003:

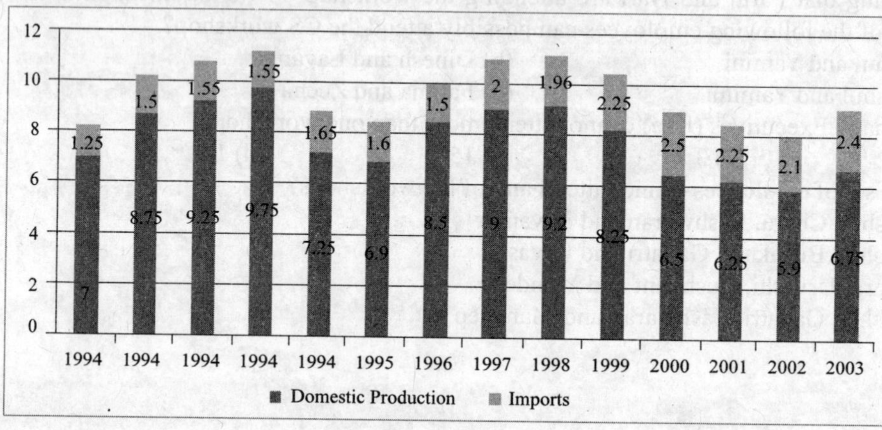

Following pie chart gives the total sales data of 2003 company wise: (100% = 9.16 lacs)

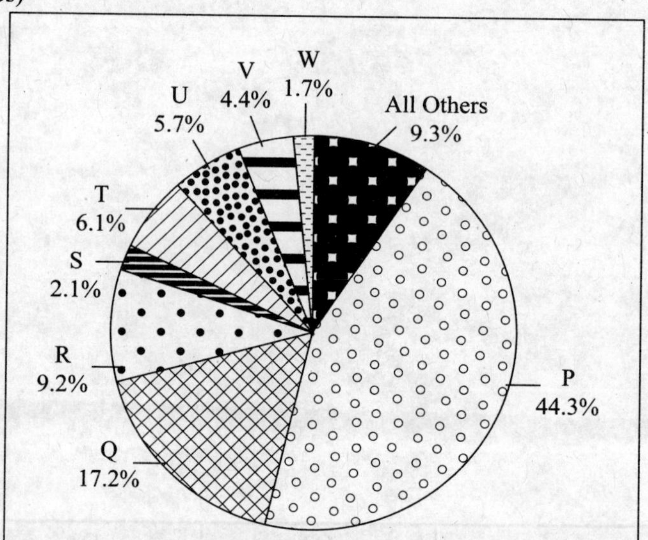

Companies P, Q, R and S are Indian companies, and only their production is domestic production. Companies T, U and V are Japanese companies, whereas company W is an Italian company. All the cars sold by T, U, W and V are imported in India.

Following pie chart provides the same data (of the year 2003) as per the country of production:

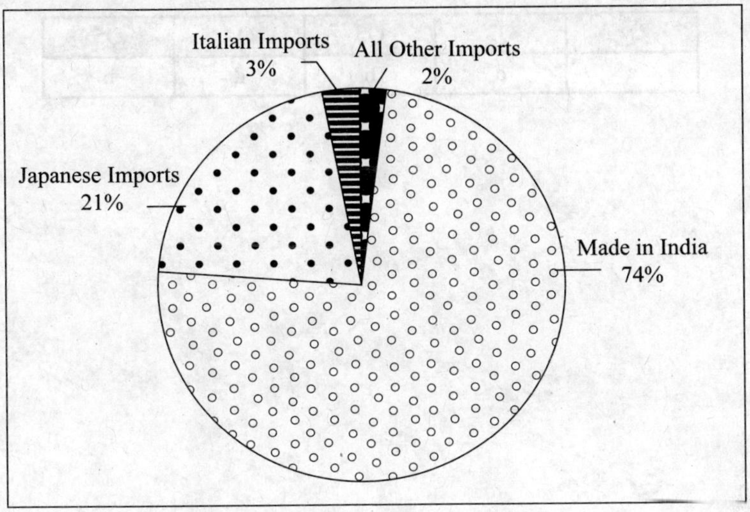

Since ALL the percentage values are approximated to fit in the data set, data sets may provide 1%–2% error in the calculations. All the answers have been approximated in such a way to handle 1%–2% margin.

Q1. What was the last year prior to 2003 in which retail sales of new cars in the United States exceeded the retail sales of new cars in 2003?

(a) 1979 (b) 1978 (c) 1977 (d) 1976

Q2. What is the number of cars sold by company T in 2004?

(a) 32000 (b) 64000 (c) 128000 (d) Cannot be determined

Q3. Approximately what percentage of all the new cars sold retail in India in 2003 was imported from Japanese manufacturers other than companies T, U and V?

(a) 2.9% (b) 4.6% (c) 6.9% (d) 9.3%

Q4. Approximately what percentage of the new domestic cars sold in India in 2003 was manufactured by Company Q?

(a) 10% (b) 12% (c) 15% (d) 23%

Q5. Approximately how many of the new cars sold in India in 2003 was imported from German manufacturers other than Company W?

(a) 32,000 (b) 119,000 (c) 156,000 (d) 192,000

Rough Work

Answer Grid

Q1	Q2	Q3	Q4	Q5
a	d	b	d	b

Missing Data

Set 82

Direction for questions 1–4

Go through the data set given below and solve the questions based on it.

Following table gives the marks of six students A1, A2, A3, A4, A5, and A6 in five papers of an examination - P1, P2, P3, P4, and P5.

	P1	P2	P3	P4	P5	Total
A1	97	99	98		99	491
A2	92	95		94		
A3	86	89		90	93	443
A4	78		76		85	405
A5		69	64	75		349
A6	49	53	48	60	55	
Total	468		463	502	505	

Q1. What is the marks obtained by A2 in P5?
(a) 98 (b) 96 (c) 94 (d) 95

Q2. What is the marks scored by A4 in P4?
(a) 80 (b) 88 (c) 85 (d) 75

Q3. What is the average of the total marks scored by all six students?
(a) 304 (b) 436 (c) 336 (d) 404

Q4. What is the sum of the marks scored by all the six students in P2?
(a) 576 (b) 486 (c) 586 (d) 476

Rough Work

■ Answer Grid

Q1	Q2	Q3	Q4
a	c	d	b

Unit Sales and Profit Set 83

Direction for questions 1–5

Go through the data set given below and solve the questions based on it.

Answer the questions based on the following two graphs, assuming that there is no fixed cost component and all the units produced are sold in the same year.

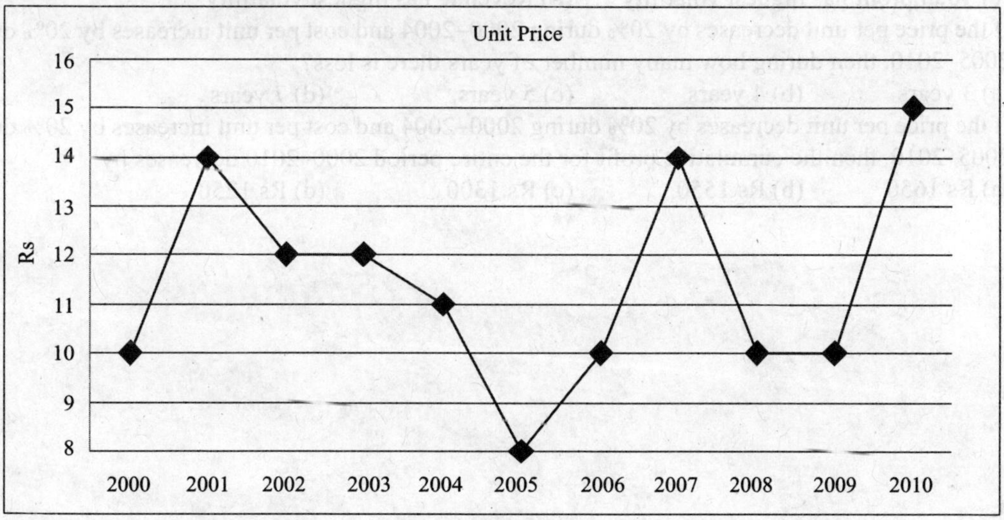

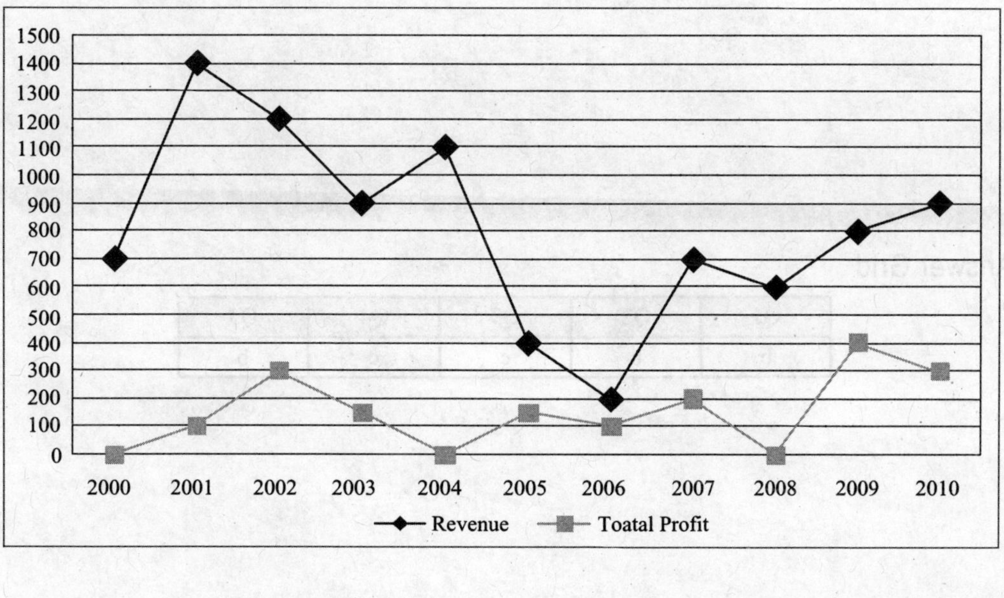

Q1. In which year per unit cost is HIGHEST?
(a) 2002 (b) 2001 (c) 2005 (d) 2007

Q2. What is the approximate average quantity sold during the period 2000–2010?
(a) 64 units (b) 70 units (c) 77 units (d) 81 units

Q3. If volatility of a variable during 2000–2010 is defined as $\frac{\textit{Maximum Value} - \textit{Minimum Value}}{\textit{Average Value}}$, then which of the following is TRUE?
(a) Price per unit has highest volatility (b) Cost per unit has highest volatility
(c) Total profit has highest volatility (d) Revenue has highest volatility

Q4. If the price per unit decreases by 20% during 2000–2004 and cost per unit increases by 20% during 2005–2010, then during how many number of years there is loss?
(a) 3 years (b) 4 years (c) 5 years (d) 7 years

Q5. If the price per unit decreases by 20% during 2000–2004 and cost per unit increases by 20% during 2005–2010, then the cumulative profit for the entire period 2000–2010 decreases by:
(a) Rs 1650 (b) Rs 1550 (c) Rs 1300 (d) Rs 1250

Rough Work

Answer Grid

Q1	Q2	Q3	Q4	Q5
b	b	c	c	b

Company Performance Set 84

Direction for questions 1–4

Go through the data set given below and solve the questions based on it.

Following bar charts provides the sales value (in Rs lac) and profit (in Rs lac) of a company over the years.

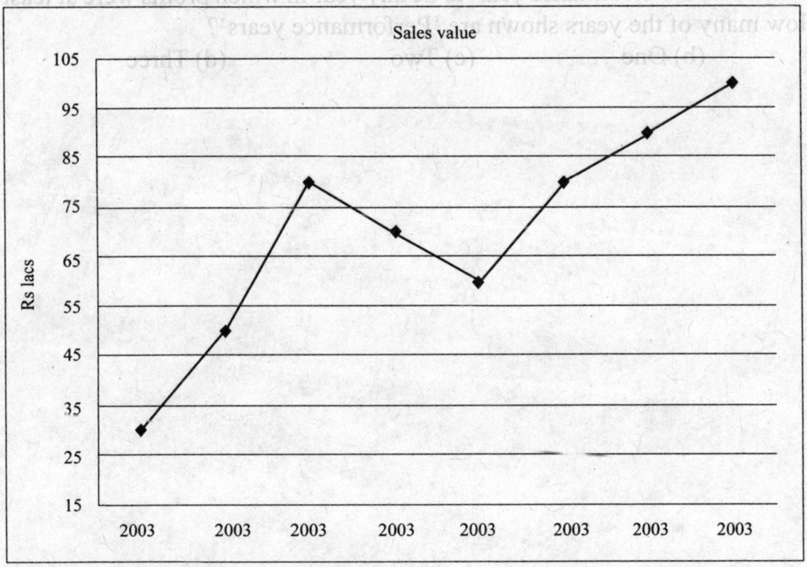

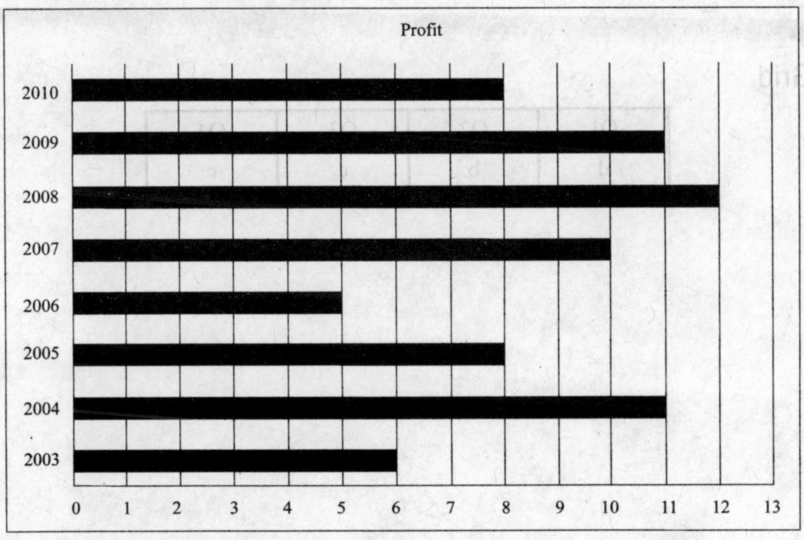

Q1. For the years 2004 to 2009 inclusive, what was the amount of the greatest increase (in Rs Thousands) in sales from one year to the next?
(a) 50 (b) 500 (c) 30 (d) 300

Q2. For the period from 2003 to 2007 inclusive, what were the average sales per year (in Rs Thousands)?
(a) 570 (b) 580 (c) 600 (d) 610

Q3. In which of the years from 2004 to 2009 inclusive, did profits change by the greatest percentage over the previous year?
(a) 2004 (b) 2005 (c) 2007 (d) 2008

Q4. Company considers a 'Performance year' to be any year in which profits were at least 20 percentage of sales. How many of the years shown are 'Performance years'?
(a) Zero (b) One (c) Two (d) Three

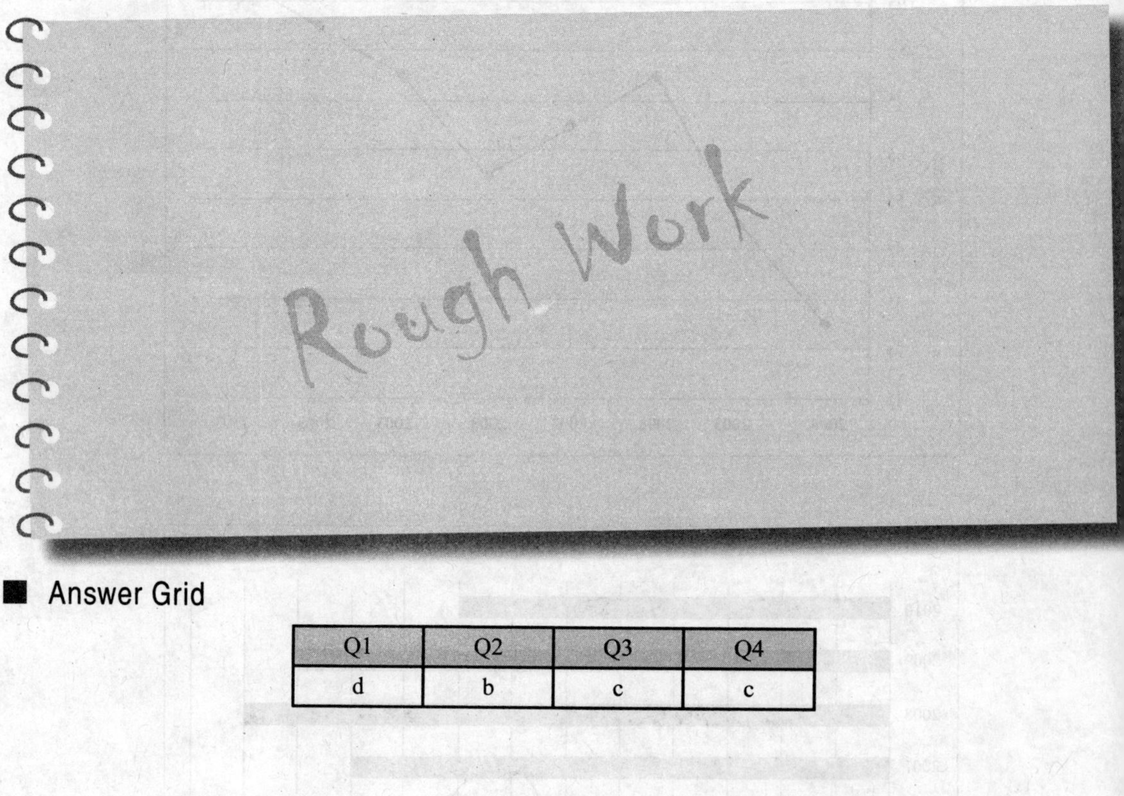

Answer Grid

Q1	Q2	Q3	Q4
d	b	c	c

Sylvan Inc. Set 85

Direction for questions 1–4

Following bar charts provide the sales revenue and profit of Sylvan Inc. for the years 1998 – 2003. Company has two divisions – Electronic and Non Electronic.

Total profit (in Rs millions)

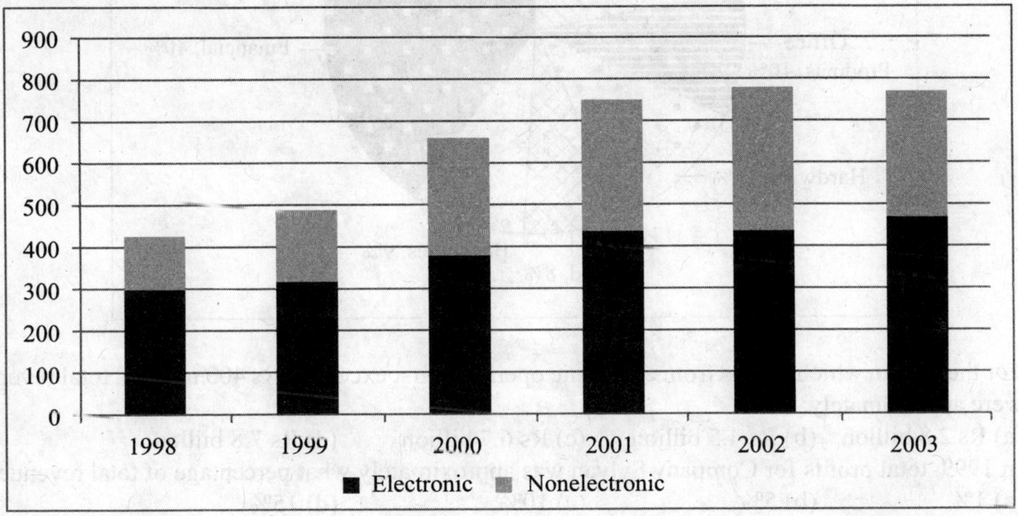

Total Sales Revenue (in Rs billions)

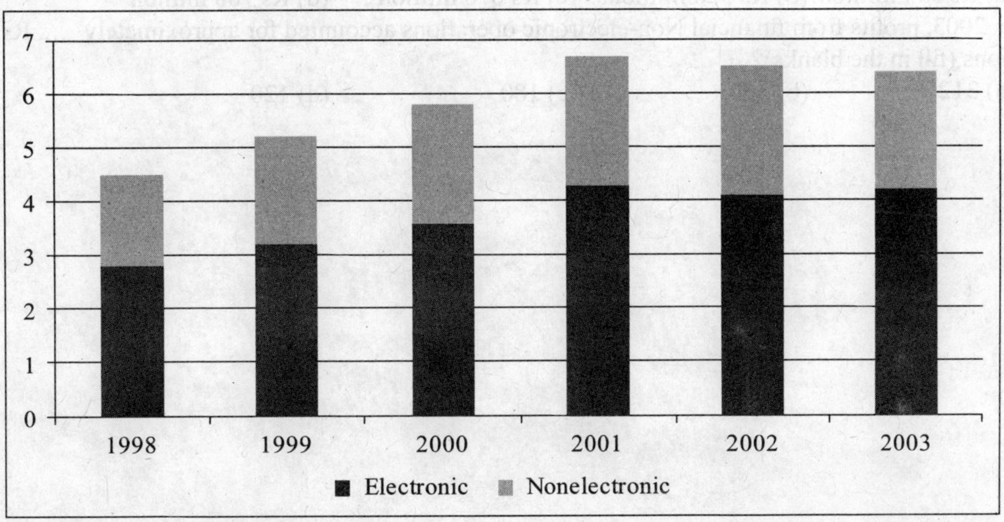

Distribution of profit (in million of Rs) from Non-electronic Operations in the year 2003 is given in the pie chart:

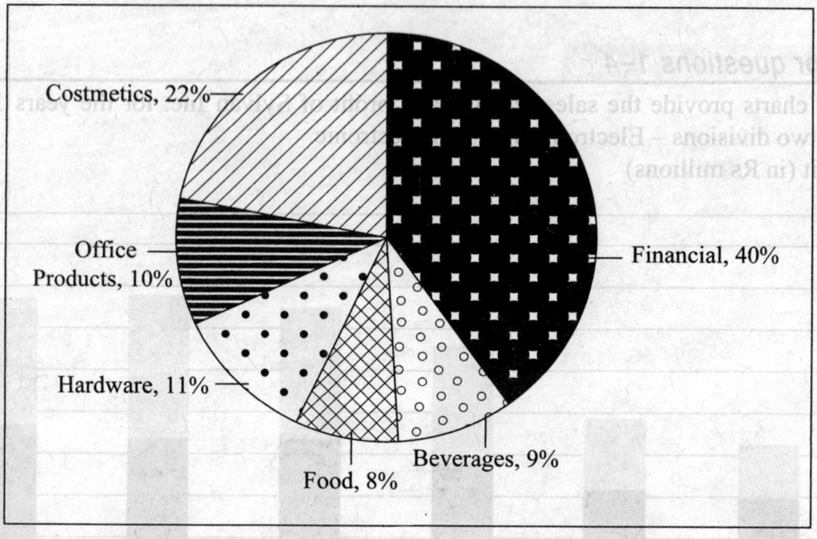

Q1. For the year in which profits from electronic operation first exceeded Rs 400 million, total revenues were approximately:
(a) Rs 2.8 billion (b) Rs 4.5 billion (c) Rs 6.7 billion (d) Rs 7.8 billion

Q2. In 1999, total profits for Company Sylvan was approximately what percentage of total revenues?
(a) 1% (b) 5% (c) 10% (d) 15%

Q3. For the two years in which profits from electronic operations were most nearly equal, the combined profits from Non-electronic operations were most nearly
(a) Rs 340 million (b) Rs 520 million (c) Rs 670 million (d) Rs 780 million

Q4. In 2003, profits from financial Non-electronic operations accounted for approximately ….. Rs millions (fill in the blanks)?
(a) 312 (b) 300 (c) 180 (d) 120

Rough Work

Answer Grid

Q1	Q2	Q3	Q4
c	c	c	d